Behaviormetrics: Quantitative Approaches to Human Behavior

Volume 10

Series Editor

Akinori Okada, Professor Emeritus, Rikkyo University, Tokyo, Japan

This series covers in their entirety the elements of behaviormetrics, a term that encompasses all quantitative approaches of research to disclose and understand human behavior in the broadest sense. The term includes the concept, theory, model, algorithm, method, and application of quantitative approaches from theoretical or conceptual studies to empirical or practical application studies to comprehend human behavior. The Behaviormetrics series deals with a wide range of topics of data analysis and of developing new models, algorithms, and methods to analyze these data.

The characteristics featured in the series have four aspects. The first is the variety of the methods utilized in data analysis and a newly developed method that includes not only standard or general statistical methods or psychometric methods traditionally used in data analysis, but also includes cluster analysis, multidimensional scaling, machine learning, corresponding analysis, biplot, network analysis and graph theory, conjoint measurement, biclustering, visualization, and data and web mining. The second aspect is the variety of types of data including ranking, categorical, preference, functional, angle, contextual, nominal, multi-mode multi-way, contextual, continuous, discrete, high-dimensional, and sparse data. The third comprises the varied procedures by which the data are collected: by survey, experiment, sensor devices, and purchase records, and other means. The fourth aspect of the Behaviormetrics series is the diversity of fields from which the data are derived, including marketing and consumer behavior, sociology, psychology, education, archaeology, medicine, economics, political and policy science, cognitive science, public administration, pharmacy, engineering, urban planning, agriculture and forestry science, and brain science.

In essence, the purpose of this series is to describe the new horizons opening up in behaviormetrics — approaches to understanding and disclosing human behaviors both in the analyses of diverse data by a wide range of methods and in the development of new methods to analyze these data.

More information about this series at http://www.springer.com/series/16001

Ryozo Yoshino

Cultural Manifold Analysis on National Character

Methodology of Cross-National and Longitudinal Survey

Springer

Ryozo Yoshino
Department of Culture and Information Science
Doshisha University
Kyotanabe-shi, Kyoto-fu, Japan

ISSN 2524-4027 ISSN 2524-4035 (electronic)
Behaviormetrics: Quantitative Approaches to Human Behavior
ISBN 978-981-16-1672-3 ISBN 978-981-16-1673-0 (eBook)
https://doi.org/10.1007/978-981-16-1673-0

This Springer imprint is published by the registered company Springer Nature Singapore Pte Ltd.
The registered company address is: 152 Beach Road, #21-01/04 Gateway East, Singapore 189721, Singapore

Preface

Cross-National Comparative Survey is a treasure trove of social surveys.

(Chikio Hayashi, [1], Chap.5. Translated by the present author).

The late Chikio Hayashi (1918–2002), a leading figure of Japanese statistics since the postwar period, once mentioned this after a long experience of survey research. As a key member of the Institute of Statistical Mathematics (ISM), he has initiated and developed a longitudinal and international comparative survey by ISM for more than half a century. The present author has been a member of the survey team for the past three decades.

In this book, I will explain about our past surveys and a paradigm called *Cultural Manifold Analysis (CULMAN)* that have been developed for the longitudinal and cross-national comparative research under the statistical philosophy of "Data Science."

Some explanation is needed here to distinguish the "Data Science" from the "data science" recently used in the fields, such as AI, IT, or big data technologies. This term was coined by Hayashi in the 1980s. At a keynote speech by the International Classification Society (IFCS) held in Kobe in 1996, Hayashi explained that conventional hypothesis testing, numerical models, and statistical models were not suitable to the study of complex and ambiguous phenomena such as human science and social science. And he proposed to construct "Data Science" based on a data-driven, exploratory, and holistic approach that deals with such complex and ambiguous phenomena (Hayashi, 1998a; Osumi, 2003).

In the postwar period (1945–1955), the members of ISM worked for reorganizing statistical systems of the government and civil sections, including mass media polls and marketing research in order to recover national economy and develop postwar democracy in the devasted country. Among various works, Hayashi started the Japanese National Character Survey (JNCS) with his colleagues, Hiroshi Midzuno, Hirojiro Aoyama, and Shigeki Nishihira in 1953. This survey has lasted now for more than 60 years. To my knowledge, this is the longest lasting statistically random sampling survey of people's consciousness. And Hayashi and Nishihira expanded it

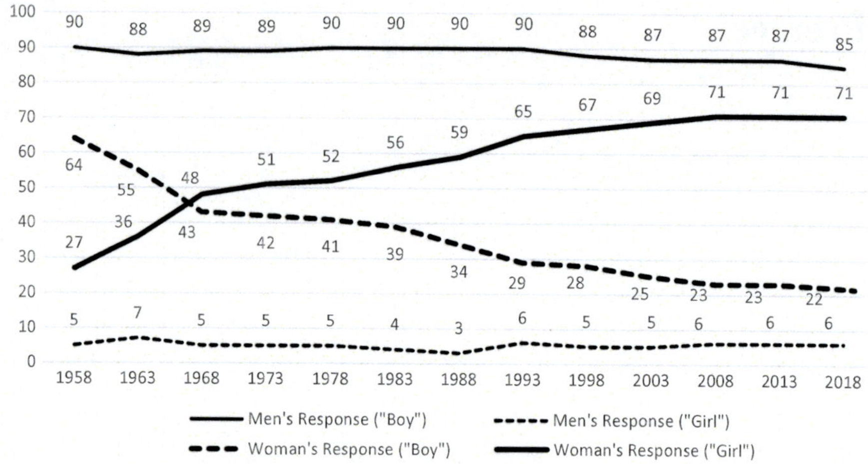

Fig. A "If you could be born again, would you like to be born as a man or a woman?" (Japanese Data).

into cross-national research in the 1970s, when Japan was close to the peak of rapid industrial development. Meanwhile, Hayashi had been the leader of the Japan Statistical Society that contributed to the cabinet policymaking of several prime ministers, and served as presidents of many national and international academic communities. Not surprisingly, all these experiences are closely linked to the birth of the "Data Science" (Hayashi [2, 3], which is expected to contribute mainly to the practical solution of various social problems.

Scientific measurements can be thought of as the interaction between the observer, the observed object, and the representation of the measurement system (i.e., as models, theories, statistical graphs, and tables). No matter how good a model or theory is, it cannot be considered a reality in itself. It's just one of the possible expressions of reality. The model or theory is the "finger pointing to the moon" in *Zen* teaching. The finger itself is not the "moon," but it helps others understand what and where the moon is. We have created *Cultural Manifold Analysis (CULMAN)* as a "finger" for others to understand our survey research.

JNCS includes dozens of questions. Here is an example of question in JNCS. "If you could be born again, would you like to be born as a man or a woman?" Fig. A shows the stability of Japanese men's response "man" (some 90%) as well as the significant change of the Japanese women's response from "man" to "woman," in the past 60 years. This question is directly related to the issue of gender equality. But among other questions, it may seem to symbolize the steady change in social situations since the postwar period in Japan (see Chap. 2 for details). However, before jumping to any particular interpretation on the distribution of response data, it is better to understand the holistic changes in the social systems, social values, and international relations behind the survey data. For reference, see Figs. B1 and B2 of the Asia-Pacific Values Survey. Countries seem to exhibit their own modalities closely related to their history and politics.

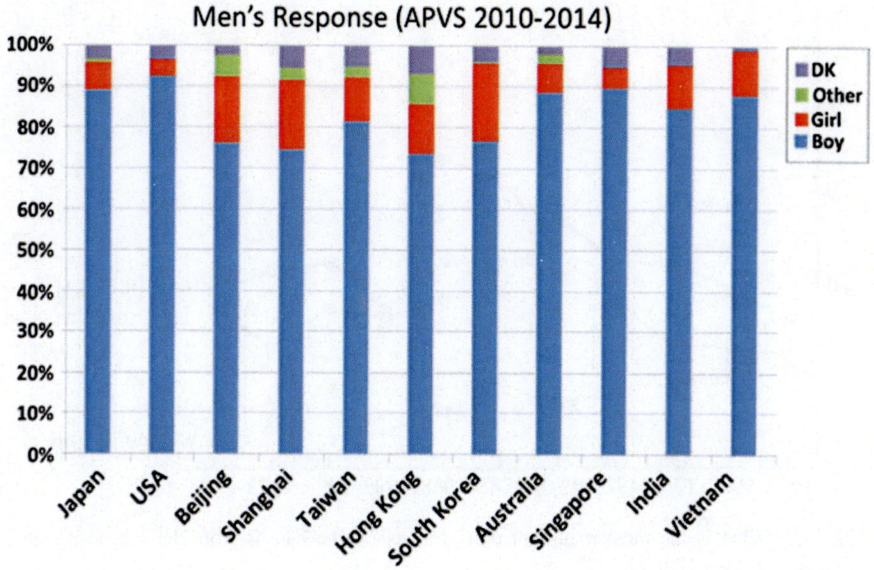

Fig. B1 Men's response in the Asia-Pacific Values Survey to the question "If you were born again, would you like to be born asa a boy or a girl?"

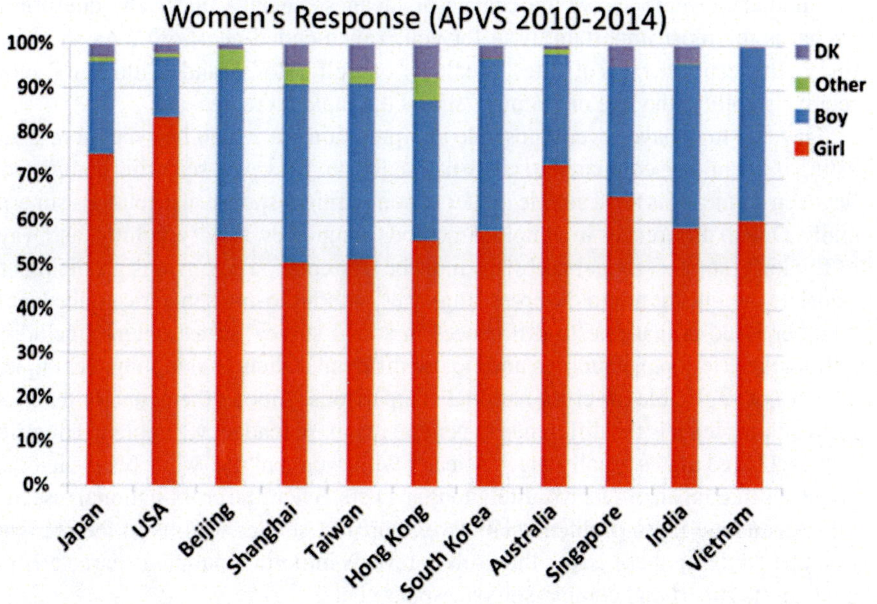

Fig. B2 Women's response in the Asia-Pacific Values Survey to the question "If you were born again, would you like to be born as a boy or a girl?"

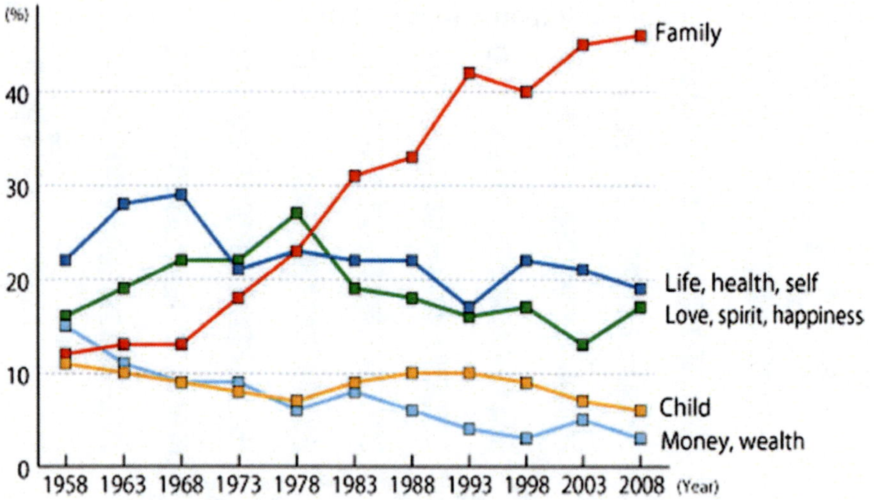

Fig. C "What is the most important thing for you?" The data is from the Japanese National Character Survey (JNCS).

Another example shows the change of Japanese social values. The question is "What is the most important thing for you? (open-ended question)." As shown in Fig. C, the response rates of "life," "health," "oneself," "love," and "child (ren)" show relative stability; the rate of "family" shows the rapid increase.

The data may have several possible interpretations. It would be even more interesting if it could be compared to cross-national data. However, scientific comparison between countries is not a simple matter of comparing response distributions superficially. Due to differences in sampling methods, languages, history, culture, and ways of thinking, some essential problems must be overcome. The obvious problem is to translate the question into different languages without losing semantic equivalence. Another issue is that due to differences in social survey infrastructure (including political restrictions), countries need to use different statistical sampling techniques. This is not negligible for cross-national comparisons. Among them, one of the most difficult problems is the difference in general response tendency. People in countries like the United States usually answer clearly with "yes" or "no," while people in other countries like Japan often give ambiguous answers. Scientific cross-national research must overcome these problems. However, our past studies lead us to the recognition that studying these issues themselves reveals important national characteristics beyond the superficial comparison of response data.

The world in 2020 began with the coronavirus turmoil from China. Differences in government behavior, media coverage, scientist activity, and people's panics clearly reflect differences in values and attitudes toward life, science, and the political system.

Over the past 65 years, our research team has collected statistical random sampling data of people's awareness and opinions with development of our paradigm to justify longitudinal and cross-national research. The purpose of our research is to promote mutual understanding of people's attitudes, behavioral manners, religion, and values. Although our work is still under construction, an overview of past work may give the reader some ideas for further development. I am a survey statistician with a background in mathematical psychology and psychometrics, but I'm not an expert in anthropology, sociology, or international relations. Therefore, I may need to refrain from developing substantive discussions for those empirical sciences. However, I believe that experts in a variety of disciplines, such as economics, politics, religion, sociology, and international relations, can use our data to find new ways to advance their own research.

The ancient Chinese martial artist Sunji said, "If you know the enemy and you know yourself, you are not in danger of 100 battles. It is best to win without fighting." To the people of other countries, the Japanese attitude may often seem vague. This may be derived from the principles that are deeply ingrained in the Japanese (self-restraint of behavior, self-restraint of freedom). Mutual understanding is the key to avoiding unnecessary conflicts between countries. To that end, it is important to strive to make other countries understand our values and ideas.

I hope this book will help expedite empirical social science and evidence-based policymaking for world peace and prosperity.

Kanagawa-ken, Japan Ryozo Yoshino

References

1. Hayashi, C. (1984a). Science of survey (in Japanese). Tokyo: Kodan-sha. (Reprint by Chikuma Academic Bunko in 2001).
2. Hayashi, C. (1998). The quantitative study of national character: interchronological and international perspectives. In Sasaki, M. (ed.), Values and attitudes across nations and time, (pp. 91–114). Boston: Brill.
3. Hayashi, C. (2001). Science of data (in Japanese). Tokyo: Asakura-syoten.

Acknowledgements

I would like to thank the many collaborators who have guided me in many surveys and carried out the surveys with me over the past thirty years. I am deeply grateful to Akinori Okada for giving me the opportunity to publish this book. I would like to thank Chikio Hayashi, Tatsuzo Suzuki, Fumi Hayashi, Kazue Yamaoka, Masamichi Sasaki, Yoshiyuki Sakamoto, and Zheng Yuejun for developing the survey research.

I am deeply grateful to my advisors when I was a graduate student at the University of California, Irvine. Tarow Indow, William H. Batchelder, Louis E. Narens, Robert Duncan Luce, Donald D. Hoffman, and Bruce M. Bennett influenced me to develop the philosophy on the methods of empirical science, theory construction, and measurement theory in mathematical psychology. My old friends, Paul Shirey, Scott Simpson, Nancy Mear, Vincent Brown, Frank Pollick, Gregory B. Pollock, Kimberly A. Jameson, and Eri Tatekawa also supported me.

I cordially appreciate Mr. Yutaka Hirachi, Ms. Shinko Mimura, Mr. Sridevi Purushothaman, and Mr. Gowtham Characravarthy at Springer Nature for their assistance for publishing this volume.

This book is dedicated to my parents, Mitsu Yoshino and Kinjiro Yoshino, and to my brother Kotaro Yoshino.

Note: This book introduces various survey data collected by the Institute of Statistical Mathematics (ISM). For details of the questionnaire, survey method, and total of each survey, please refer to the series of ISM Survey Reports and related websites.

1. http://www.ism.ac.jp/editsec/kenripo/index.html (ISM Survey Research Reports in Japanese).
2. http://www.ism.ac.jp/editsec/kenripo/index_e.html (ISM Survey Research Reports in English).
3. http://www.ism.ac.jp/ism_info_j/kokuminsei.html (JNCS in Japanese).
4. http://www.ism.ac.jp/ism_info_e/kokuminsei_e.html (JNCS in English).
5. http://www.ism.ac.jp/~yoshino/ap2/index.htm (ISM Cross-National Studies in Japanese).
6. http://www.ism.ac.jp/~yoshino/index_e.html (ISM Cross-National Studies in English).

Among the cross-national data, the following data are frequently used in this book.

7. The Seven Country Survey (1987–1993) (Hayashi et al., 1998).
   https://www.ism.ac.jp/editsec/kenripo/contents_e.html (ISM Survey Report, No.63, 64, 70–73, 75–82).
   https://www.ism.ac.jp/~yoshino/arito/eg/top_e.htm (Website 1).
   https://www.ism.ac.jp/~yoshino/data/1_e.html (Website 2)
8. The East Asian Value Survey (EAVS) (1992–1995) (Yoshino 2006).
   https://www.ism.ac.jp/~yoshino/sr/index.html (Summary Report).
   https://www.ism.ac.jp/~yoshino/ea/index_e.html (Website).
9. The Pacific-Rim Value Survey (PRVS) (2004–2009) (Yoshino 2010).
   https://www.ism.ac.jp/~yoshino/sr/index.html (Summary Report.)
   https://www.ism.ac.jp/~yoshino/pr/index_e.html (Website).
10. The Asia-Pacific Value Survey (APVS) (2010–2014) (Yoshino, Shibai & Nikaido, 2015).
   https://www.ism.ac.jp/editsec/kenripo/pdf/kenripo117.pdf (Summary Report).
   https://www.ism.ac.jp/~yoshino/ap2/index_e.html (Website).

For a detailed list of publications

11. https://www.ism.ac.jp/~yoshino/references_e.html.

Contents

Chapter 1
History of the Japanese National Character Survey (JNCS)

All survey studies are woven by the trinity of history, and theory and practical methodology of the research, and should not be described only by superficial data. Therefore, I start this book by a brief history ([1]).

Most of today's statistical sampling survey systems for Japanese official and private sectors have been established as joint research by experts from various fields, led by the Institute of Statistical Mathematics in the decade after World War II. Those experts gathered to work for postwar national economic recovery and peace.

1.1 Postwar Democracy: From Statistics for War to Statistics for Peace

1.1.1 Birth of "Statistical Mathematics": Philosophy of Statistics

"Statistical mathematics" is a philosophy that began in the field of Japanese statistics since the early 1940s. Traditional "mathematical statistics" rely on the mathematical assumptions of probability distribution theory which are almost impossible to verify directly. Several groups of statisticians criticized this and aimed to develop a new statistical approach to solving social problems in a practical way [2]. It was named "Statistical Mathemtics." Since then, this idea has successively developed as a "Quantification Method" in the 1950s and 1960s, "Multidimensional Data Analysis," "Behaviormetrics," and "Science of Survey" in the 1970s and 1980s, and "Science of Data" since the 1990s [3–7].

Unfortunately, this new philosophy of statistics was once incorporated into war in the course of the times. In 1943, during the war, the Academic Research Council submitted a proposal the establishment of a new research institute with a focus

R. Yoshino, *Cultural Manifold Analysis on National Character*,
Behaviormetrics: Quantitative Approaches to Human Behavior 10,
https://doi.org/10.1007/978-981-16-1673-0_1

on statistical mathematics. After the Cabinet's decision in 1944, the Institute of Statistical Mathematics, under the control of the Minister of Education, took charge of the study of mathematics and its applications related to probability and promoted the research. Soichi Kakeya, Professor at Tokyo Imperial University at the time, served the post of Director-general.

It was the era when universities and research institutes all over Japan were incorporated into a wartime system called total war. For my generation born after the war, the details of what the Institute did at the time are almost unknown. But it is said that they were engaged in cryptanalysis or what is now called operations research [8, 9].

1.1.2 New Mission for Democracy

1.1.2.1 Reorganization of Official Statistics

Under the US occupation after the war, the institute was reorganized into the 1st Research Department (fundamental theory), the 2nd Research Department (statistical theories for natural sciences), and the 3rd Research Department (statistical theories for social sciences) in April 1946. This reorganization was related to Rice Statistics Mission's recommendation from the United States in December 1946.

ISM members reportedly believed that the institute created for strategic research would be abolished under the postwar occupation. Therefore, when the people from GHQ/SCAP (Headquarters General, Allied Supreme Commander) arrived, they thought the institute would soon be abolished (Hiroshi Midzuno 1991 [Personal Communication]). However, the United States had a detailed understanding of human resources in various areas of Japan, including information from interrogating Japanese prisoners of war [10]. And they used Japanese experts from various fields to create a new democracy and restore the economy and industry. Thus, the institute was given a new mission to take the lead in reorganizing official (governmental) statistics and in establishing a method of scientific opinion polling as the basis for the development of postwar democracy.

Chikio Hayashi, later a director-general of ISM but a then young member, was dispatched with Hiroshi Midzuno to the Ministry of Finance as a member of the Institute of Statistical Mathematics. There they handled a variety of large data. Midzuno was a senior to Hayashi in the Department of Mathematics at the University of Tokyo. And he inspired his intimate life-long friend Hayashi to develop new statistics.

Through this experience at the Ministry of Finance, Hayashi and Midzuno confirmed that real data could not always be on normal distribution even if the sample size of the data is large. This can be an important basis for criticizing the armchair theory of mathematical statistics. Then, in the processes of solving real-world problems such as market research, personnel management, and criminal law issues, Hayashi collaborated with Midzuno to build "Hayashi's Quantification Theory." The most striking feature of the theory is that it does not assume a priori probability distribution of variables. This frees us from mathematical assumptions that cannot be checked directly [11–14, 15].

Incidentally, after the war, Hayashi obtained the US military OR (operations research) textbook "Methods of Operations Research" [16]. And he found that Americans had thought the same in their military operational studies and had reached similar conclusions as the Japanese [17].

In this way, Japanese researchers took off military uniforms, began reorganization of government statistics, developed statistical marketing surveys for economic recovery for the reconstruction of destroyed land, and established a scientific opinion poll system for democracy. On the other hand, the US government, ending the occupation in 1953, believed that a reliable and rigorous statistical system had been established so that they could constantly monitor Japan from Washington and respond immediately in the event of an emergency [18]. In those days, the US government did not seem to believe democratization of the Japanese. Besides, in the start of Cold War, they needed to monitor people's movements including residents near the US military bases in Japan. During the Vietnam War, the US military bases faced severe opposition from the residents.

1.1.2.2 Japanese Literacy Survey

In the history of postwar surveys, it is important to explain about the "Japanese Literacy Survey" as the first nationwide statistical random sampling.

In policymaking during the US occupation, there was some argument in the US government and military that Japanese use of esoteric *kanji* (Japanese version of Chinese characters) prevents them from developing their academic abilities and democracy. On this matter, in 1948, a nationwide statistical random survey was conducted to investigate the literacy of Japanese people. The result turned out that Japanese people had sufficient literacy to develop democracy. Thus, the Japanese language was saved [19, 20].

"Romanization" (abolishment of "*kanji*" and use of alphabet) has been strongly insisted by Major Robert King Hall of the GHQ Civic Information and Education Agency (CIE) without any coordination with the US government. He seems to have conceived this idea through his own difficult experiences of learning Japanese at the time of Princeton University and the Naval Academy. However, Gordon T. Bowles, Advisor to the Education Mission of the State Department, believed that language reforms should be left to Japan and should not be enforced externally. And Hall was left out of this matter. After all, the results were in agreement with the report from Tokyo Imperial University Educational System Research Committee to President Shigeru Nambara [21, 22]. In fact, there was a complicated background, such as censorship under the occupation and conflicting views within the US government and the military. However, at least on the surface, it provided a model for policymaking based on a scientific survey of stakeholder intent.

In addition, this provided a model to check statistical random sampling theory and its practice. Prior to this survey, an exhaustive survey (almost 100% valid return rate from the target population of more than 20,000) was conducted in some cities such as Odawara City of Kanagawa Prefecture. Because each city of Japan has almost complete list of the residents (then a list of Food Distribution Books made for governmental control of distribution of food during World War II), so they used the list for a certain statistical random sample from the total population (stratified multistage sampling). Next, the distribution of literacy scores across the population was compared to the estimated distribution of scores obtained from statistical random samples. The estimated score was close to the true score within the theoretical sampling error calculated according to the sample size (e.g., 1000 or 3000 respondents from the total population). After this confirmation, this sampling method was extended to apply to the nationwide survey (for more details, see Yomikaki-nouryoku-chosa-iinnkai [20]).

1.1.2.3 Establishment of Public Opinion Poll

The method invented for the Japanese Literacy Survey provided a model for later rigorous statistical surveys, including resident survey, marketing survey, public poll survey, surveys to forecast election outcomes, etc. (Later, they began to use a list of registered residents or voters rather than a list of food distribution books). Of course, the governmental surveys started to use the method.

Usually an ordinary public poll cannot be directly checked as to whether it is conducted statistically adequate or not. Therefore, it is important to stick to a rigorous statistical method validated by theory and practice, such as the Japanese Literacy Survey. But an exception that we can directly check validity of a survey method is on prediction of election outcome based on public poll.

During the occupation, anthropologist John Perzel worked for the CIE (Civil Information Education Section), which was in charge of public opinion polling in Japan. In 1947, he was surprised to know that a Japanese survey company, Yoron Kagaku Kyokai (Public Opinion Scientific Research Center [POSRC]), successfully predicted the winner of the highly competitive election of the Governor of Tokyo metropolitan city by a statistical random sampling survey. The sample size was only 500 respondents. He knew American mass media's failures of public opinion polls on elections such as the 1938 presidential election. Therefore, he hardly could believe the validity of the Japanese survey on such a small sample size. In those days, most surveys of major Japanese newspapers collected a sample of more than 60,000, sometimes more than 200,000 respondents. He ordered the POSRC to explain about the details of the survey. Eventually, however, he understood the validity of rigorous statistical random sampling survey and admired them on the performance.

Meanwhile, Japanese researchers were surprised to learn that a US survey firm mispredicted the winner in the 1948 US presidential election. After studying American survey methods (quota sampling with respect to age, gender, and race), people found that Japanese sampling methods were much more rigorous in statistics.

The survey research of Japan was under supervision of the United States, but the Japanese successfully performed beyond the US research. However, the Japanese know that the United States, along with Japanese staff, made significant contributions to Japan during the early days of occupation, not only for survey research but also in realizing an ideal system of social security and medical care that could hardly have been realized in its own country.

Since the success of literacy surveys or the election predictive survey, statistical random sampling has become more and more dominant in surveys by governments, mass media, marketing researchers, academic researchers, and others.

1.1.2.4 Resident Surveys Under the 1945–1952 Occupation by the United States

It may be worthwhile writing about some resident surveys in islands such as Amami-Oshima Island and Okinawa Island in the occupation time. With Japanese cultural anthropologists, Hiroshi Midzuno (a member of ISM) was engaged in the survey of Amami-Oshima (Hiroshi Midzuno private communication 1991) [23]. The United States did not brief them the real purpose of the survey, but totally relied on credible Japanese statisticians and anthropologists to conduct the survey. The report was classified and sent to Professor Ishino Iwao, Ohio University, for analysis for their own purpose. As a result of the survey, the residents of Amami had an overwhelming sense of belonging to Japan, but the residents of Okinawa did not. (It may be natural given cruel actions of the Japanese military to Okinawa residents during the war). It was the time that complicated international relations were seen, such as the resident movement and the start of Cold War between the United States and the Soviet Unions. After all, Amami-Oshima returned to Japan several years after the war, whereas the American military base was built in Okinawa and it had been under the US custody until 1972 [23, 24].

The United States entered the Cold War era and asked many Japanese anthropologists to survey islands throughout Japan [25]. The nine academic societies such as anthropology, psychology, and sociology were organized as an association for survey research and they carried out many surveys [26]. This can be thought of as a model of the basic spirit of intelligence: independence between the information collector and the information analyst. It can also be said to be a model for policymaking based on empirical research data collected by highly credible local researchers.

After the war, the United States immediately disarmed Japan and demanded to enact a peaceful constitution. Japan peacefully fulfilled its demands. However, the Soviet Union and China began targeting the disarmed Japan. Then, the United States changed its policy and demanded Japanese rearmament in anticipation of the Korean War [27]. It became the origin of the still lasting subordination of the Japanese government and the Japanese Self-Defense Force to the United States Military [28, 29].

Many documents on survey research during the US occupation were declassified around 1990 and returned to Japan Association for Public Opinion Research (JAPOR). The documents are currently kept by Waseda University.

1.2 The Japanese National Character Survey (JNCS) Since 1953

1.2.1 History and Theory

In 1953, the Institute of Statistical Mathematics started the "Japanese National Character Survey," based on the statistical random sampling method developed practically in the "Japanese Literacy Survey" [7, 2]. The word "national character" may be a problem in academics and politics, but it is used as a nickname for our research. In our terminology, "national character" refers to a characteristic that is reflected in the response patterns of people's consciousness surveys in a country. This is closely related to Inkeles' [31] concept of "national character" regarding the statistical mode of responses of people. Incidentally, as for measurements of national character, Inkeles [31] claimed that aspects directly related to economic or political conditions should not be regarded as part of the "national character." It is reasonable, however, to assume that people of different countries may respond differently to certain economic or political items on a questionnaire even under the same economic and political conditions, and that such differences in response patterns may be closely related to "national character." For example, in the late 1980s, Brazilians showed a high degree of life satisfaction and happiness even when their country was experiencing severe conditions regarding international debt. In contrast, the Japanese did not show a high degree of life satisfaction and happiness even when their economy was close to being the best-performing one in the world (cf. Easterlin paradox).

Today, this survey is called one of the three major statistical sampling consciousness surveys in Japan, together with the "Shakai-ishiki ni kansuru Chosa [Social Survey on People's Consciousness]" by the Cabinet Office and the "Nihon-jin no Ishiki Chosa [Japanese Consciousness Survey]" by NHK [Japan Broadcasting Corporation]. The team of JNCS thought that "national character" wouldn't change greatly over years; therefore, they didn't necessarily assume to repeat the survey over years at first. And many new question items were introduced in the second survey of 1958, five years after the first survey. However, they found considerable changes in the response distributions for some of the same questions, recognizing the importance of conducting a longitudinal survey with the same question items. However, from the 8th in 1988, taking into account changes in the times, they created two types of questionnaires: K-type (continuous type) mainly consisting of the same questions and M-type (future-oriented type) which replaced some items with those to capture possible future changes of Japanese attitudes.

This survey, which has been continuing for over 65 years [7, 30, 32–34], (See the appendix written by Yoshino [31]), is a globally unique longitudinal survey supported by the Ministry of Education. It has subsequently prompted similar types of statistical surveys in other countries such as ALLBUS (Germany), CREDOC (France), Eurobarometer, the European Values Survey, and the General Social Survey (GSS) in the United States. Meanwhile, many experimental research on survey methodologies have been conducted by ISM since the 1950s.

1.2.1.1 Kameda's Simple Random Sampling Theory

Regarding the theory of statistical random sampling, Toyojiro Kameda presented a theory of simple random sampling and applied it to calculate the accuracy of sampling data on the first Japanese Census Data (1920) in 1923 (see Note 1). Furthermore, he applied the theory to the work of security insurance, and also applied to the estimation of damage caused by the Great Kanto Earthquake in Tokyo and its surrounding areas in 1923. He reported the results at the 19th Conference of the International Statistics Institution (ISI) held in Japan in 1930 [35, 36]. Although it was the era of "Taisho democracy," a peak of democracy before World War II, his theory had never been applied to opinion polls. (Taisho era is 1912–1926. Democracy movements peaked in the 1910s and 1920s).

However, in the 1948 Japanese Literacy Survey, Japanese staff did not pay attention to Kameda's work. They studied American books on statistics borrowed from the CIE library and invented a practical method of nationwide random sampling suitable for Japan. As mentioned, officially registered "food distribution books" were used to design multistage stratified sampling. (In later public opinion surveys, they started to use list of resident registration or voters that are updated regularly and are almost complete. Compared to other countries, this gives Japan a significant advantage in the precise estimation of sampling error of public opinion survey).

1.2.1.2 Multistage Stratified Sampling of JNCS

This section describes the sampling design of Japanese National Character Survey (JNCS). The basic design has been the same, but the detailed steps have changed slightly over the years. For more information, please refer to the homepages of Japanese National Character Survey Committee (https://www.ism.ac.jp/kokuminsei/en/page9/page10/index.html), or Mizuno, et al. [30]. The following is from the explanation of the 13th survey [37].

In general, the procedure for the 13th survey was the same as for the 8th to 12th surveys. The selection of samples for each survey was performed by a three-stage stratified stochastic sampling method. First, we stratified boroughs, cities, towns, and villages, considering population size and demographic variables. One administrative area was randomly selected from each level so that the probability of selection was proportional to the size of the population. In the 13th survey, 400 districts were selected (Stage 1). Second, from each randomly selected district, a small area called CHO/AZA (the same as the census unit) was selected in the same way (Stage 2). Finally, respondents were selected from the resident register of the selected district using systematic random sampling (Stage 3). In the 13th survey, a total of 6,400 respondents were selected.

In the 12th and 13th surveys, we refined the geographic stratification and increased the number of sampling points and sample size (number of respondents) compared to the 8th to 11th surveys. We also used the resident listings in the last two surveys, with some exceptions, instead of voters' listing previously used. Thus, the sampling point unit was a voting district in the previous surveys, but "CHO/AZA" in the recent surveys.

The surveys up to 8th one were conducted with the cooperation of many universities. In these surveys, districts were assigned to universities, and members of the research committee visited the universities and directed student interviewers. However, since the 9th survey in 1993, each survey has been conducted with the cooperation of a private survey company.

In the case of the 13th survey, prior to the fieldwork survey, the sampling manager randomly selected respondents from the resident registry according to the designated procedures at the city hall. An average of 16 respondents were selected for each sampling district. Fieldwork was conducted from the last 10 days of October to the first 10 days of December 2013. As a result, 3,170 (1,591 for the K-type and 1,579 for the M-type) were completed from the target sample of 6,400, and the completion rate was 50%. Details of incompletion rates are summarized in Tables 1 through 5 of Nakamura et al. [37].

1.2.1.3 Installment of JNCS

In the 13th survey, the installment of the survey was entrusted to the Nippon Research Center (NRC) (Gallup International Association). However, prior to the interview, the Institute of Statistical Mathematics (ISM) processed some of the respondent sampling from the registered resident card (80 out of 400 locations in Tokyo, Osaka, and two other prefectures). The researchers of NRC and ISM responsible for sampling visited the town hall, ward office, and city hall of their town or village and sampled there from the Resident Register using a designated statistical random sampling method.

We then mailed a request form of cooperation to the sampled respondents in advance. The interviewer visited each house, met with the respondents directly, handed over the request from the Institute of Statistical Mathematics, and requested cooperation. If respondents were unavailable due to absence or other reasons, a new date and time was set for the interviewer to visit again.

When a respondent agreed to cooperate with the survey, the interviewer read the questions on the questionnaire, and recorded the answers provided by the respondent verbatim on the questionnaire. For questions indicated with show cards of response choices, the cards were presented to the respondents. If a sampled respondent didn't agree to cooperate, the interviewer recorded it as "refusal." If the interviewer was unable to meet the respondent after visiting many times for some reason, the reason was recorded according to each situation, such as "temporary absence," "long absence," or "moved."

Note

Note 1. For simple random sampling, the estimated sampling error$E$ at a 95% confidence interval for the percentage of a statistics in the population (e.g., the response percentage of "yes") is given mathematically as.

$$E = \pm 1.96\sqrt{(P(1-P))/n(N-n)/(N-1)},$$

where P is the response percentage in the population ($0.0 \leq P \leq 1.0$), N is the population size, and n is the sample size. For large N (e.g., over 10,000), this is approximately equal to.

$$E \doteqdot \pm 1.96\sqrt{(P(1-P))/n}.$$

If n is large enough, P is approximated by the observed response percentage p in the sample.

$$E \doteqdot \pm 1.96\sqrt{(p(1-p))/n}.$$

This is maximized when p = 0.5. Often, the estimated sampling error is treated roughly as

$$E \doteqdot \pm 1/\sqrt{n}.$$

For example, if the sample size n is 10,000, the sampling error of the estimated response percentage is ± about 1%. If the sample size n is 1,000, the sampling error of the estimated response percentage is ± about 3%.

In practice, you should also consider non-sampling errors (a variety of errors, including errors in sampling coverage of population, recording responses, processing data, etc). Non-sampling error cannot be calculated accurately in theory, but it is sometimes assumed to be approximately at the level of sampling error [38].

In the early days when modern computers were unavailable, accurate calculation on a set of big data was very time-consuming and costly. As a result, even if a large set of data was generated by an exhaustive survey, a tentative rough calculation was often done using a small dataset randomly sampled from the original large set of data. Some researchers insisted that calculating statistics on random sampling made sense only if exhaustive survey data were available. From the point of view of modern statistics, it seems strange. It took a long time before random sampling surveys like today became frequently used.

1.3 Summary

After World War II (late 1940s to 1950s), the government and private sector statistical systems and institutions were intensively reorganized as the basis for regaining national power under Japan's new democratic regime. In 1953, the Institute of Statistical Mathematics (ISM) launched the "Japanese National Character Survey" based on the statistically rigorous sampling method devised in the 1948 "Literacy Survey of Japan." This survey is based on multilayered and multistage statistical random sampling from a nearly complete list of voters or residents across the country.

JNCS has been in operation for over 65 years. It is also currently known as one of the three major longitudinal statistical consciousness surveys, together with the "Shakai-ishiki ni kansuru Chosa [Social Survey on People's Consciousness]" by the Cabinet Office and the "Nihon-jin no Ishiki Chosa [Japanese Consciousness Survey]" by NHK [Japan Broadcasting Corporation]. The survey motivated other countries to launch similar longitudinal general surveys, such as the General Social Survey (GSS) in the United States, the Eurobarometer in the EU, and ALLUBUS in Germany.

JNCS was closely linked to the reorganization of official statistics and the establishment of statistical public opinion polls to develop Japan's postwar democracy. It also symbolizes the development of Japanese statistical philosophy such as "Statistical Mathematics," "Quantification method," "Multidimensional Data Analysis," "Behaviormetrics," and "Science of Data."

References

1. Yoshino, R. (in press). "Japanese National Character Survey" to "Cross-National Comparative Survey on People's Consciousness"–From "Statistical Mathematics" to "Science of Data." Special issue of the 14th Japanese National Character Survey. In Proceedings of ISM.
2. Midzuno, H. (2003). What is "Statistical mathematics"? (in Japanese). *Japanese Journal of Behaviormetrics, 30*(2), 191–192.
3. Hayashi, C. (1998). The quantitative study of national character: Interchronological and international perspectives. In M. Sasaki (ed.), *Values and Attitude Across Nations and Time,* (pp. 91–114). Boston: Brill.
4. Hayashi, C. (2001). *De-ta no kagaku (Science of Data).* Asakura-syoten: Tokyo.
5. Yoshino, R. (2001). *Kokoro wo hakaru (Measuring the mind).* Tokyo: Asakura-syoten.
6. Yoshino, R., & Hayashi, C. (2002). An overview of cultural link analysis of national character. *Behaviormetrika, 29,* 125–142.
7. Yoshino, R., Hayashi, F., & Yamaoka, K. (2010). *Kokusai- hikaku deta no kaiseki (Analysis of comparative survey data).* Tokyo: Asakura-shoten.
8. Inose, N. (1983). *Lost war of 1941 summer* (in Japanese). Tokyo: Sekai-Bunka-sha. (Republished by Bunsyun Bunko in 1986 and by Chuo Bunko in 2010).
9. Kimura, H. (2002). Wartime research by Japanese mathematicians in WWII (in Japanese). RIMS *RIMS Kokyuroku* (Kyoto University), 1257, 260–274.
10. Nakata, S. (2010). *Interrogation center, P.O. Box 651, Tracy, California (in Japanese).* Kodansha: Tokyo.
11. Hayashi, C. (1993). *Quantification method (in Japanese).* Tokyo: Asakura shoten.
12. Midzuno, H., Hayashi, C., & Aoyama, H. (1953). *Quantification and prediction—a study on prediction of election outcomes (in Japanese).* Tokyo: Maruzen Co.
13. Morimoto, E. (2005). The development of statistics in Japan after the second world war: The formation and dissemination of Hayashi's quantification methods (HQM). *Japanese Journal of Behaviormetrics, 32*(1), 45–67.
14. Morimoto, E. (2012). The formation of quantification methods (HQM) (in Japanese). *Advances in Social Research, 9,* 5–16.
15. Takahashi, M. (2002). Philosophy of Hayashi's quantification method: A study from a viewpoint of survey researcher (in Japanese). *Shakai-To-Chosa (Advances in Social Research), 9,* 17–23.

16. Morse, P. M., & Kimball, G. E. (1946). *Methods of operations research.* U.S. Navy (A Revised version published by The Technology Press of MIT in 1951, and further republished as the Dover edition in 2003).
17. Maruyama, K. (2015). *Life of Chikio Hayashi—creation of data science (in Japanese).* Tokyo: Shinyou-sha.
18. Kondo, Y. (1953). *Statistical analysis of agriculture of Japan* (in Japanese).
19. Hidano, T. (2018). Reports of the United States Education Mission to Japan and Reform of Japanese Language and Script : A role of psychologists (in Japanese). *Shinrigaku-shi · Shinrigaku-ron (History of Psychology and Psychology Studies),* 18–19, 1–14.
20. Yomi-kaki-nouryoku-tyousa-iinnkai (Committee of Japanese Literacy Survey). (1951). *Japanese literacy survey (in Japanese).* Tokyo University Press.
21. Kayashima, A. (2000). *A study of alphabetization of Japanese language: Analysis of domestic and international factors.* Tokyo: Kazama-shobo.
22. Unger, J. M. (1996). *Literacy and script reform in occupation Japan.* Oxford University Press.
23. Eldridge, R. D. (2003). *Amami return and US-Japan relations (in Japanese).* Tokyo: Nanpo-shinsha.
24. Takahashi, M. (2002). Residential survey in Amami Island under the occupation by the U.S.: A study on Bennett's report (in Japanese). *Intelligence, 6,* 92–98.
25. The Japanese Society of Ethnology. (1952). Special number: Social research—Symposium on the history and prospects of social research in Japan. *The Japanese Journal of Ethnology, 17*(1), 1–99.
26. Sakano, T. (2012). Conjoint surveys by the nine academic society and homeland. *The Industrial Management Project Report "Industrial Promotion and Local Communities"*, 35, 1–15.
27. Ezaki, M. (2019). *Korean War and operations to invade Japan and Taiwan (in Japanese).* Tokyo: PHP-Sinsho.
28. Suenami, Y. (2012). *The identity of Japan's subordination to the US-report from the US Archives (in Japanese).* Tokyo: Kobun-sha.
29. Tanigawa, K., & Sudo, Y. (2019). *Japanese Origin of slavery to the United States: Reading "1959 US Confidential Document" (in Japanese).* Tokyo: Ohtsuki-shoten.
30. Mizuno, K., Suzuki, T., Sakamoto, Y. Murakami, M., Nakamura, T., Yoshino, R., Hayashi, C., Nishihira, Y., & Hayashi, F. (1992). *Dai 5 Nihon-jin no Kokuminsei* (The fifth volume of Japanese National Character Survey). Tokyo: Idemitsu-syoten.
31. Inkeles, A. (1997). *National character.* New Brunswick: Transaction Publishers.
32. Hayashi, C. (1992). Quantitative social research-belief systems, the way of thinking and sentiments of five nations. *Behaviormetrika, 19*(2), 127–170.
33. Hayashi, C. (1992b). Belief systems, Japanese way of thinking: Interchronological and international perspectives. *Social, educational and clinical psychology, Proceedings of the 22nd International Congress of Applied Psychology,* **3,** 3–34. Hove (HK): Lawrence Erlbaum Associates, Publishers.
34. Sakamoto, Y. (2000). How has the Japanese way of thinking changed? A half century of the statistical survey of the Japanese national character. *Proceedings of the Institute of Statistical Mathematics, 48*(1), 3–32.
35. Kawasaki, S. (2020). Roles of official statistics in the historical development of statistics in Japan. *Journal of the Japan Statistical Society, 49*(2), 161–186.
36. Takahashi, M. (2004). History of science and a scientist–*Interview with Chikio Hayashi* (in Japanese). *Japanese Journal of Behaviormetrics, 31*(2), 107–124.
37. Nakamura, T., Yoshino, R., Maeda, T., Inagaki, Y., & Shibai, K. (2017). A Study of the Japanese National Character: The Thirteenth Nationwide Survey (2013) - English Edition. *ISM Survey Research Report,* No. 119. The Institute of Statistical Mathematics.
38. Hayashi, C. (ed.) (1984a). *Science of survey* (in Japanese). Tokyo: Kodann-sha. (Reprint by Chikuma Academic Bunko in 2011).

Chapter 2
Findings of the Japanese National Character Survey (JNCS)

2.1 Start of the Longitudinal Survey

Initially, the JNCS research team did not intend JNCS as a longitudinal survey because they expected the "national character" to be stable over the long term. Thus, the second survey in 1958 introduced many items on various topics that were not covered in the first survey in 1953. However, the results turned out to show significant differences on some of the same items used in both surveys. This observation led the JNCS team to recognize the importance and necessity of conducting a longitudinal survey of the same items. Thus, except for some items, the same items were used as in the first survey from the third survey until the fourth survey in 1968.

However, around 1970, Japan's remarkable social change required new items to capture new aspects of the Japanese people. So, they launched two types of questionnaires: K-type ("Keizoku" [continuous]) and M-type ("Mirai" [future]). The K-type consists of almost the same items as a series of surveys, but the M-type captures new aspects by replacing some items of the K-type with new items. This idea later leads to the research paradigm of cultural link analysis (CLA) as it expands to cross-national research in the 1980s. (Unfortunately, the core concepts of K-type and M-type were not necessarily well maintained, and the distinction between these questionnaire items was confused in later surveys. Chikio Hayashi had been the leader of JNCS but retired from the Institute of Statistical Mathematics in 1988. After retirement from ISM, he began to focus on cross-national survey. Therefore, the remaining members of the Japanese survey may not have fully recognized the importance of the distinction of questionnaires in survey design).

This section summarizes the main findings of JNCS. To understand the findings, it is helpful to take into account historical background and generational features in the Japanese character.

There are many ways to divide the period from after World War II to the present. The following classification is an example (see Table 2.1).

R. Yoshino, *Cultural Manifold Analysis on National Character*,
Behaviormetrics: Quantitative Approaches to Human Behavior 10,
https://doi.org/10.1007/978-981-16-1673-0_2

Table 2.1 Moden History of Japan (Showa, Heisei and Reiwa Periods)

Era	Period
Showa Era (1926–1988)	
1927 Showa Financial Depression	Prewar Period
1928 Zhang Sakulin bombing incident	
1931 Manchurian Incident	
1932 5.15 case (a military coup of May 15)	
1933 Withdraw from the League of Nations	
1936 2.26 case (a military coup of February 26)	
1937 The Sino-Japanese War began	
1939 World War II began	
1940 Japan-Germany-Italy Military Alliance	
1941 The Pacific War (Greater East Asia War) began	
1945 August: Atomic bombs dropped on Hiroshima and Nagasaki. Acceptance of Potsdam Declaration September: Signing of surrender document	GHQ Occupation Period (1945–1952)
1946 Enactment of the New Constitution of Japan (enforced in May 1947)	
Showa era (after the war)	
1950 Farmland reform	
1951 the San Francisco Peace Treaty and the (former) Japan-US Security Treaty	
1952 Sovereignty restored	Postwar Recovery and Rapid Economic Developmen (1952–1973)
1956 Joined the United Nations	
1960 (New) Japan-US Security Treaty entered into force Intense student movement against the treaty	
1964 Tokyo Olympics held	
1971 Dollar shock (Nixon Shock)((unilateral cancellation of the direct international convertibility of the US dollar to gold)	
1973 Oil Shock (the second oil shock in 1979)	Recession and Stable Development
Heisei Era (1989–2019)	Bubbling Economy (1986–1991)
1989 Emperor Showa passed away (Showa 64/1989) Enthronement of Emperor Heisei First introduction of Consumption tax enforcement (3%)	

(continued)

Table 2.1 (continued)

Era	Period
1993 The End of "1955-political system"	Recession after the End of Cold War
1995 The Great Hanshin-Awaji Earthquake Subway Sarin Case by Cult Aum-Shinrikyo	
2009–2012 Goverment by Democratic Party of Japan	
2011 The Great East Japan Earthquake and Fukushima-Daiichi Nuclear Power Plant Accident	
2012 Return to Government by Liberal Democratic Party	
Reiwa Era (2019)	
2019 Emperor Heisei Yields the Throne Enthronement of Emperor Reiwa	
2020 Global Coronavirus Turmoil	

1. GHQ Occupation Period (1945–1952)
2. Postwar Recovery and Rapid Economic Development (1952–1973)
3. Recession and Stable Development (1973–1986)
4. Bubbling Economy (1986–1991)
5. Recession after the End of Cold War (1991–present).

(After collapse of Bubbling Economy).

Table 2.2 shows a generational classification of the Japanese. Each generation has a nickname and a unique character. (Of course, this does not mean that everyone in a particular generation has the dominant character of his/her generation). These tables will help you understand the findings of JNCS survey.

A series of books on JNCS was published under the name of JNCS committee in 1961, 1970, 1975, 1982, and 1992) [1–5], semi-regularly until 1992 as a complete summary of research and analysis. However, no single full-text summary containing recent results has been published (except for popular reading such as Hayashi and Sakuraba [6].

Now we have 65 years accumulated data of JNCS, so we can see that some of earlier observations or predictions have turned out to be true and some others wrong. (For details of tabulations of data and related information, see the following websites of JNCS.

https://www.ism.ac.jp/kokuminsei/en/index_e.html ... The Website of JNCS.

https://www.ism.ac.jp/editsec/kenripo/contents_e.html ... The Pdf of Printed Survey Research Reports

Table 2.2 Generation Nicknames of the Modern Japanese

Generation name	Birth year	Age at 2018	Characteristic
WWII Generation	up to 1928	90 years or older	・Educated **before or during the wartime** Conservative on politics, family life, gender role, etc.
Post-war First Generation	1929–1943	75–89 years old	・Born before or during the wartime, but many educated in **postwar democracy**. Realized the **rapid economic development** after the war
Baby Boomers	1944–1953	65–74 years old	・Educated in postwar democracy, witnessing the rapid economic development as well as the negative aspects such as environmental pollutions. Experienced the **student movement** of the 1960s. Politically liberal, some people are radical
New Humankind	1954–1968	50–64 years old	・Grown up in the rapid economic development, watching TV. **New attitdes and values** on family life, gender role, religion, etc.
Baby Boomer Junior	1969–1983	35–49 years old	・Grown up in various media such as video games and PCs. Witnessed the end of the Cold War. The **collapse of the bubble economy** in the 1990s has created a difficult job market
New Human Junior	1984–2002	16–34 years old	・Many grew up in an age of "**yutori-kyoiku**(relaxed education)" and a prolonged economic depression. Cell phones and computer networks are now available. Their attitude and values are similar to those of the baby boomer junior

Source This is from "Contemporary Japanese Consciousness Structure" (NHK BOOKS 1260) (NHK Broadcasting, Institute for Cultural Studies, 2020, pp. 218–228), summarized by the present author

https://www.ism.ac.jp/editsec/kenripo/pdf/kenripo119.pdf … The Report of the 13th JNCS.

https://www.ism.ac.jp/~yoshino/references_e.html) … Reference of Books and Papers.

2.2 Early Analysis on Surveys 1953–1978

This period includes the period immediately after the end of World War II and the period of rapid economic growth. In 1956, the white paper of Ministry of Finance declared that "it is no longer the time of postwar." The author intended to warn of a possible reduction in various supports by the United States and other countries regarding the economic restoration of Japan. But most Japanese took it in the meaning of "we've regained economic power in peace." Still, at that time many Japanese were in poverty. However, they were gradually shifting industrialization from agriculture to commerce and industry, first in light industry and then in heavy industry. Later we realized that we were in a period of rapid growth. Average yearly growth rate was more than 10% between 1955 and 1973. Then we witnessed the dollar shock in 1971 (the Nixon shock regarding the unilateral cancellation of the direct convertibility from the US dollar to gold), the first oil shock in 1973, and the second oil shock in 1978.

As for this period of JNCS, Hayashi and Suzuki [7], two of the key members of the project provided a brief but comprehensive summary of the changing Japanese belief system, quality of life issues, and 25 years of social conditions. Also, their paper emphasizes the importance of multidimensional analysis such as "Hayashi's Quantification Method III (QMIII)" (a sort of principle component analysis for categorical data, equivalent to correspondence analysis by Benzecri [8]), dual scaling by Nishisato [9], or optimal scaling [10]. Experience with past survey analyses has shown that simple cross-tabulations and simple breakdowns between two or three questions do not always reveal changes in belief systems and attitudes. However, the effective application of QM III has revealed the consistency and character of change over the 25 years in postwar Japan.

The main findings of Hayashi and Suzuki [7] can be summarized as follows.

First, since the beginning of the Meiji era (1868), Japanese people have maintained an emotional conflict between Japanese tradition and modernization (Westernization) even after World War II before around 1978 (the sixth survey in 1978). In the life of enjoyment, individualism, self-confidence, politics, and institutions, there has been a major change from traditional to modernization, perhaps through education and mass media. However, some Japanese traditional attitudes remain unchanged, especially concerning human relations ("giri-ninjyo" attitude), religiosity, and basic ways of thinking. Therefore, in the process of accepting Western things with the Japanese spirit and emotions, it seemed that both the modernization of Japanese culture and the Japanization of Western culture were mixed.

Second, but then, the 6th survey (1978) showed signs of the birth of a new system of beliefs, indicating a turning from modernization to tradition. These signs appeared in the young group (aged 20–24) born about 10 years after World War II. In 1956, the Economic White Paper declared that "Japan is no longer postwar." For many Japanese, this symbolized the beginning of industrial and economic development. However, around 1970, Japan faced serious pollution problems as a result of its rapid industrialization.

It was not a simple regression to "tradition." This change suggested that the pattern of change in Japanese belief system was significantly different from the trend that persisted until the early 1970s. Afterwards, we came to know that it had been the time for Japan to transition from rapid economic growth to a more stable economy. Economic development approached the top of the world in the 1980s. In the 1980s, those young people were named "new mankind" because the elderly could hardly understand their attitudes and values.

In subsequent developments of research, we gradually came to learn the deeper meaning of those findings. For example, human or interpersonal relationships have been stable in each country for decades, but each country may have shown each own style of relationships. "*Giri-ninjyo*" is one of the major characteristics of human relations in Japan. "Giri-ninjyo" can be loosely defined as attitude and affection for traditional Japanese relationships. "*Giri*" is a Japanese value roughly corresponding to duty, obligation, or even burden of obligation. "*Ninjyo*" is the human affection that inescapably springs up in conflict with social obligation. Within the Japanese worldview of "giri-ninjyo," people sometimes feel serious conflicts between "giri" and "ninjyo," and often cannot make quick decisions.

Several items in JNCS were aimed at measuring "giri-ninjyo" attitudes and were more or less successful up to a certain point of time. However, today people rarely use the phrase "giri-ninjyo" in daily life, so those items may have become obsolete, although this does not necessarily mean that the Japanese have lost the spirit of "giri-ninjyo." Therefore, nowadays it seems necessary to invent new items to capture the spirit of "giri-ninjyo." Longitudinal surveys assume that the same item will be used for years to maintain comparability. However, if the survey series lasts more than a few decades, items may need to be changed or replaced appropriately for long-term comparability. This is similar to the long-term measurement of commodity price changes in economics. In economics, the average price index is calculated from the average value of many daily commodities. But with the times, the commodities that should be calculated are gradually replaced to keep pace with the changes of the times, with maintaining its continuous comparability. (This idea leads to the separation of K-type and M-type questionnaires, and ultimately to the invention of the CLA paradigm that will be explained in next chapter).

Third, regarding religion, it is known that Japanese tend to have religious faiths as they grow older. On the other hand, the majority of Japanese consider religious heart/mind to be important, even though only a third of them have religious beliefs. In general, this trend has been stable, but has become somewhat ambiguous due to global secularization and due to the cohort effect of the age group born between 1935 and 1945. This age group overlaps with the generation of the student movement in the 1960s. They were born during World War II and experienced a rapid change from war to postwar. They witnessed a sudden change of school teachers from militarism to democracy. Inevitably they became skeptical of authorities, including politics and religion, and grew more "rational" in the age of postwar democracy. Subsequent research has revealed that their attitudes and values are very different from both older and younger generations. (This is a clear cohort effect appeared in the change from wartime to postwar).

Incidentally, Inglehart [11] also analyzes data from JNCS 1953–1978. He criticized the Japanese analysis of Flanagan [12] and justified his own materialist/postmaterialist theory. However, he acknowledged that the long-term trend of modernization was curvilinear rather than linear. In short, the pattern of rationalization and secularization may apply to the first stages of modernization from agriculture to urban industry to a bureaucratic society. But an advanced industrial society may bring about the reversal of the changes that have taken place in the transition from agricultural to industrial societies. This reversal seems closely related to the observations of Hayashi and Suzuki [7] mentioned above and Sakamoto [13] explained in the next section.

Here, I will not go into detail on Inglehart's modernization theory which has been gradually modified in response to extensive data accumulation over decades. However, note that the latest version of his modernization theory is presented based on evolution [14]. This is relevant to our methodology of cross-national comparison and will be discussed later.

More importantly, Inglehart [11] recognized an important issue of comparison between East and West nations. He found that items in the list of postmaterialist that were valid in the West were neutral or even reversed in Japan. This means that the scaling items need to be changed according to the context of each society in different cultures. This is also closely related to our methodology of cross-national comparison described later. Over the last few decades, our research has seen many other examples where it is not always appropriate to use the same set of items to compare countries for a particular issue, such as religion [15, 16, 17]. We need to consider scaling in cultural context.

(By the way, throughout this book, you may notice the difference between Western and Eastern research methods. Western researchers often use analytical hypothesis testing and theory construction, but Eastern researchers, or at least many Japanese survey researchers are working on holistic data-driven approaches (i.e., understanding complex reality by collecting empirical data and showing various aspects of the data without building a particular theory. Meyer [18] describes the essential national differences in the framework of thought she witnessed in international business).

2.3 Data Analysis as of 2005

After the 11th JNCS, Sakamoto [13] summarized his analysis of half-century data (1953–2003). He was a member of JNCS from the 5th survey (1973) to the 11th survey, and he worked as the director of the 9th through 11th surveys (1993–2003).

Sakamoto [13]'s analysis is summarized as follows.

(1) In the first half of 50 years, Japanese attitudes and opinions changed dramatically from the "traditional" Japanese way of thinking to "modern" ones in terms

of economy, politics, and society. On the other hand, family, schools, workplaces, and communities have seen little change in interpersonal relationships. (Sakamoto seems to use the word "modern" to mean Western style).

(2) The mid-1970s is a clear turning point separating the 50 years from 1953 to 2003. In the era of the "Nixon shock" and the "oil shock" that stopped the **rapid economic development, Japanese people showed significant changes in** attitudes, consciousness, opinions, etc. The response patterns of many items changed the long-standing trend since 1953. This was called the "return to tradition," in contrast to the trend described in (1). [13], Table 1.1, comparison between the changes of response distributions during 1953–1973, 1973–1978, and 1978–2003). However, "Return to Tradition" is a misleading term. Considering later survey data, it is more complicated than a simple return. Many of these items have not shown a steady trend of "Returning to Tradition" since 1978, but have shown a reversal or cessation of change.

(3) As for the latter half of the 50 years, the following trends are remarkable.

a. **As the first choice of the most important thing (open answer question), the** rate of "family" has been remarkably increasing since 1973 (see Fig. 2.1).
b. The response patterns of some items symbolize Japanese women's social advancement (gender equality). (See Fig. 2.3). This would be more interesting when compared to other countries (see also Figures B1 and B2 in Preface).
c. People's interest in protecting the natural environment is growing (see Fig. 2.5).
d. **There has been a clear shift in political attitude, including a significant** increase in non-partisan supporters.
e. The 10th survey (1998) showed pessimistic responses about Japanese people's self-esteem and their future, probably due to prolonged recession.

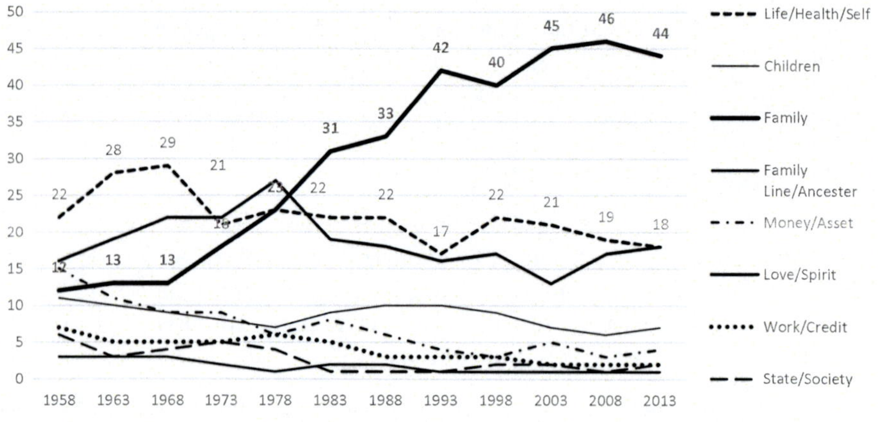

Fig. 2.1 "What's the most important thing for you?"

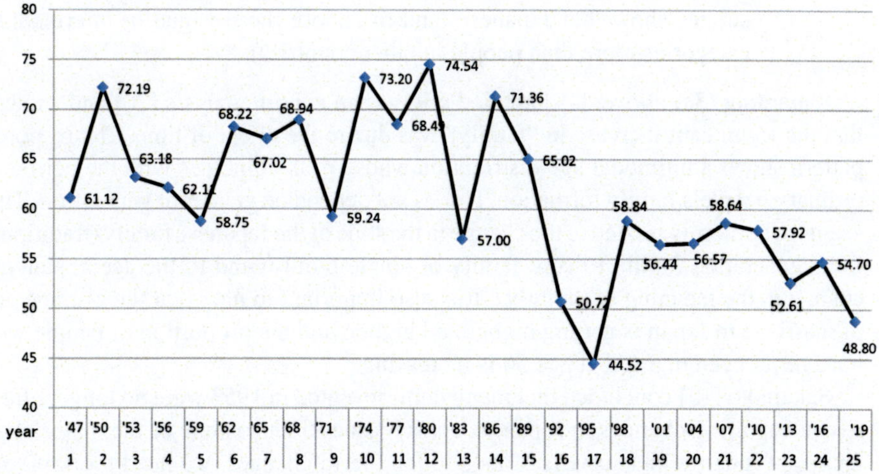

Fig. 2.2 Change of Voter Turnout of the Senator Election % in Japan (every 3 years since 1947). *Source* The Ministry of Internal Affairs and Communications. https://www.soumu.go.jp/senkyo/senkyo_s/news/sonota/ritu/ (Accessed January 31, 2020)

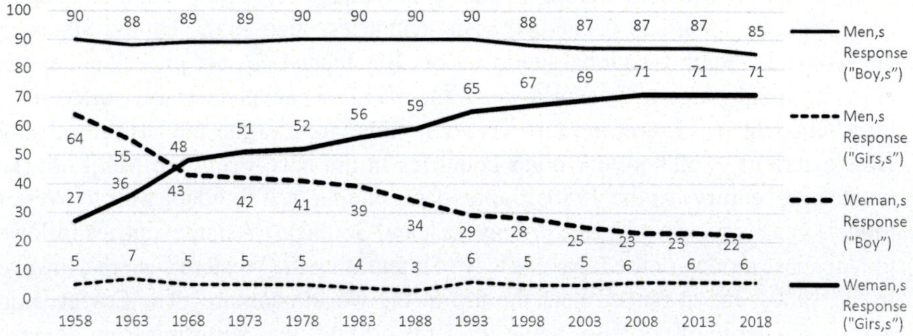

Fig. 2.3 "If you were born again, would you like to be born as a boy or a girl?"

And the same situation is still continuing ("Lost 20 years"). Japanese people's satisfaction with their lives has dropped significantly, but not so severely that they feel poverty and serious dissatisfaction.

f. Stable response patterns in interpersonal relationships and religious attitudes have been observed for many years, but the variability in response patterns for these items has increased somewhat. This may be closely related to the growing tendency to prioritize private life.

g. Recently, the response rate of ambiguous choices such as "DK (I don't know)" and the middle category (other than extreme attitudes or responses) is increasing. Sakamoto interpreted this as a result of the diversification of people's opinions and values that cannot be adequately captured by the current items. (In Sect. 3.4 of Chap. 3, cross-national studies show that Japanese tend to choose these vague or intermediate categories more than people in other countries).

Regarding (3a) above, [19] applied a Bayesian cohort analysis [20] and showed that the significant increase in "family" was due to the effect of time. The response pattern shows a unimodal age distribution whose peak coincides with the period of ordinary people's family formation. It does not depend on gender or generation. This seems to be closely related to the change in the style of the Japanese family (traditional three generation family to core family or single family) and to the accompanying change in the meaning of "family." It is also important to note that the average age of marriage in Japan is getting higher and higher, and the proportion of people who have never been married at age 50 is increasing.

Sakamoto [13] concluded that many items invented in 1953 were no longer effective at capturing changes in Japanese consciousness. This is one of the major challenges of longitudinal surveys. To study long-term changes, we need the same set of items, but items also may have a limited life span of effectiveness. Therefore, we need to invent ways to replace some items with new items while maintaining comparability over the years. (This is related to the development of paradigm of CLA or CULMAN that will be explained in next chapter).

Incidentally, attention should be paid to certain aspects of a country's characteristics regarding political decisions. Some countries, such as the United States, are aggressive in creating a global grand vision. But Japan does not present a clear and concrete grand vision to the world, and rather passively adapts to frameworks created by other countries. Therefore, Japan is often said to look vague and has caused undesirable serious conflicts with other countries in the past. From the Meiji era (late nineteenth century) to the 1980s, Japan had done its best to adapt well to Western models in a Japanese way. Later many East and Southeast Asian countries followed the Japanese model, or the Japanese way of integrating the Western system. However, in the 1980s, Japan ranked near the top of the world economy. The Soviet Union was on the verge of collapse, while the United States was suffering from domestic and trade deficits. Since then, Japan has lost the ideal model for further development and must create its own model. The government repeated substantial economic and political reforms under the name of "external pressure (among others, from the USA)," but ultimately resulted in the "lost 20 years" of the 1990s and the 2000s. All of these seem to prove that Japan is not good at leading the world according to certain practical grand plans.

As Sakamoto's review revealed in 2005, the period from 1973 to 1978 was a turning point in the postwar Japanese attitude. In 2011, however, the Great East Japan Earthquake and the subsequent Fukushima nuclear accident had a much stronger direct impact on the Japanese.

2.4 Longitudinal Data for 65 Years (1953–2018)

2.4.1 Overview

This section shows the longitudinal response patterns of several items that have been used continuously for the past 65 years. I do not show all the items here, but readers can see all the items in the homepage of ISM: https://www.ism.ac.jp/ism_info_e/kokuminsei_e.html.

Here I pick up the items that seem to show significant changes of Japan after World War II. Some data reaffirm Sakamoto's analysis of the 1953–2003 data in the previous section. Some other data show slightly different aspects that may be closely related to the damage of the Great East Japan Earthquake in 2011.

First, Table 2.3 shows the long-term trend of completion rates of JNCS, that is, the percentage of respondents who cooperated the survey and gave valid responses in the statistically selected sample. This long-term pattern is similar to the turnout pattern in the Senate elections (the Senate elections are held every three years) (See Fig. 2.2). By age and gender, older people tend to vote more than young people, and women tend to vote more than men. This also applies to the cooperation of opinion surveys. Typically, ideal statistical sampling requires a sufficiently high response rate (the cooperation rate of survey) for a valid statistical estimate. However, survey cooperation rates are closely related to public interest in the survey topic, social contribution, and participant burden [21]. If the cooperation rate is low, it is likely that people are less interested in the survey theme and social contribution to participation in the survey, and/or the burden of cooperation (time, labor, etc.) is considered to

Table 2.3 Completion Rate of JNCS (the Japanese National Character Survey) (See also Fig. 2.2.)

Survey Year	Sample size	# coompletion	Completion rate (%)
XIV (2018)	6,400	3,211	50
XIII (2013)	6,400	3,170	50
XII (2008)	6,398	3,302	52
XI (2003)	4,193	2,350	56
X (1998)	4,200	2,680	64
IX (1993)	5,400	3,738	69
VIII (1988)	6,000	3,682	61
VII (1983)	6,000	4,429	74
VI (1978)	5,400	3,945	73
V (1973)	6,055	4,594	76
IV (1968)	4,000	3,033	76
III (1963)	3,600	2,698	75
II (1958)	2,991	2,369	79
I (1953)	2,254	1,875	83

be heavy. Therefore, cooperation rates themselves represent an important aspect of people's social attitudes. The recent low response rates (cooperation rates) may indicate a decrease in people's interest in public issues (or increased interest in private life) and an increased sensitivity to privacy protection.

Table 2.4 is an updated version of Table 1.1 of [13]. The rankings correspond to the rate of decline in choices between 1953 and 1973. Higher ranking items show a rapid change between 1953 and 1973, while lower ranking items show a stable trend. You can see that the rate of change in items related to political awareness and daily life is high. The Japanese have steadily changed from state-centric to individual-centric, free and democratic politics, and from traditional families to modern nuclear families. The 1960s were a period of serious political movement, especially over the renewal of the Japan-US Security Alliance. However, in the 1970s, the first phase of "Mutou-ha-so" (voters who do not steadily support particular political parties) was recognized, and since then, those people have grown steadily and now account for the majority of voters. Therefore, for today's mass media and politicians, it is important to analyze the voting behavior of "Mu-touha-sou" to predict election results. On the other hand, the basic concepts and values of interpersonal relationships have been stable over decades.

As Sakamoto confirmed, many items showed a break in change or a reversal in the direction of change during 1973–1978. Since 1978, the response distribution for many items has tended to fluctuate up and down [13]. Thus, he concludes that these items, invented in 1953, are outdated to capture recent trends.

At point of 2020, I would think that future surveys should update or consolidate survey items on immigration and foreign students and workers, the global environment, interpersonal relationships (especially "giri-ninjyo" items), nuclear power plant concerns, etc. Many new items should be introduced. A certain item may, however, turn out to be very meaningful in special cases, even after decades of insignificant change. For example, the item on "anxiety about nuclear power plants" has been included in past our surveys (not just JNCS) since around 1988. And the anxiety of nuclear power plant accidents was much lower and stable than anxiety concerning car accidents and unemployment (#2.30 of JNCS). Therefore, it was omitted in some of our surveys in the 2000s. However, the Great East Japan Earthquake of March 11, 2011, and the subsequent disaster at the Fukushima Daiichi Nuclear Power Station, caused enormous damage to Japanese people. The disaster must have raised people's anxiety, but because it was omitted in those surveys, it was not possible to investigate direct changes immediately before and after the disaster. This means that unexpected events can regain the usefulness of items that are considered out of date. Therefore, simply stopping the change in the response rate of a survey item may not make it easy to discard it. In addition, cross-national research may add importance to items that may look completely outdated. No change in the response of certain items under certain circumstances may prove to be a unique feature of the Japanese.

Table 2.4 Change of Response Distribution (Items on Ranking of High Decreasing Rate) (Higher ranking items show a rapid change between 1953 and 1973, while lower ranking items show a steady trend. During the years 1973–1978, many items showed abortion of change or reversal of change direction. Since 1978, the response distribution for some items has tended to fluctuate up and down)

Ranking	JNCS code	1973–1953	1978–1973	2003–1978	2013–1978
1. Adption of Child (#)	#4.1	−38	−4	−14	−12
2. Prime Minister's visit to Ise Shrine (dedicated to the deity of the Imperial Family)	#3.9	−32	−5	−1	–
3. Would like to be born as a boy if one could be born again (Female's response)	#6.2	−22[a]	−1	−16	−18
4. Vote always at national elections	#8.6	−21[a]	4	−10	−8
5. To teach a young child that money is most important	#4.5	−21	0	−14	−13
(Leave our politics to an excelent leader[s])	#8.1	−20	–	–	−10
(Japanese people is inferior to the western people)	#9.6	−19	–	–	−9
6. Prefer life of clean and justice	#2.4	−18	0	−4	−6
7. Mechanization never decrease richness of human mind	#7.2	−16	11	−11	−1
8. Advancement of science and technology decreases humanity	#7.1	−13	8	−17	−13
9. Individuals become happier only when their country progresses	#7.4	−11	1	−3	−2
10. Have religious faith	#3.1	−10[a]	9	−4	−6
11. Respects one's own ancester	#4.11	−10	5	−13	–
12. Better to deny a bad rumor about one's own young childrens' teacher	#4.4	−7	−4	−12	−6
13. Give priority to the critical emergency meeting, even if your parent is in a critical condition	#5.1b	−6	3	7	−9
14. Employ the benefactor's child instead of the highest scorer	#5.1c−2	−6[b]	8	−11	−9

(continued)

Table 2.4 (continued)

Ranking	JNCS code	1973–1953	1978–1973	2003–1978	2013–1978
15. Need to conquer nature for human happiness	#2.5	−6	−1	−11	−10
16. Money and assets are the most important	#2.7	−6[a]	−3	−1	−2
17. Persist on your claim despite violating custom	#2.1	−5	−6	−9	
18. Prefer a warm-hearted boss even if he/she impose you hard works	#5.6	−4	6	−10	−10
19. Even if you need to attend a critical emergency meeting, prioritize returning to see a benefactor in a critical state	#5.1	3	0	−10	−12
20. Recruit the highest scorer, not relatives	#5.1c−1	−2[b]	−1	1	6
21. Filial piety and the respect of the individual's right	#5.14	−2[b]	0	2	−2

Source [19] or [13] Updated with recent data by the present author
[a]The Item was missing in the 1953 survey, so 1958 data was used instead
[b]The Items was missing in the 1953 and 1958 surveys, so the 1963 data was used instead
– The item was missing at least one of the surveys

The rest of this chapter presents data for some items. I do not necessarily discuss all items thoroughly, but each data will come to help highlight an important aspect of Japanese consciousness.

2.4.2 *Gender*

Figure 2.3 shows the response pattern for the following question concerning gender. The question is expressed as follows.

(#6.2) "If you were born again, would you like to be born as a boy or a girl?".

Over the last half century, Japanese men's choices of "boy" have been stable, but Japanese women's choices have changed significantly from "boy" to "girls." More precisely, about 90% of men's choices have been "boy" for 60 years. On the other hand, in 1953 women's choice of "boy" and "girl" was 26% and 62%, respectively. However, these proportions gradually changed steadily, and in 1993 the pattern was completely reversed: the percentages of "boy" and "girl" became 26% and 62% in

1993. This data seems to symbolize the steady changes in Japan's social system after the war. Also this data appears to correspond to policy initiatives and global trends at various points in the history of the enactment of the Basic Act for Gender Equality in Japan. (See the website of the Cabinet. https://www.gender.go.jp/about_danjo/law/kihon/situmu1-1.html#id2). It would be more meaningful to compare countries on this sort of data with their political reforms in recent years.

2.4.3 *Lifestyle*

JNCS include the question of "#2.4 Attitudes toward Life" as follows.

#2.4(Hand card) There are all sorts of attitudes toward life. Which one of the following statements would you say comes closest to your way of life?

1. Work hard and get rich
2. Study earnestly and make a name for yourself
3. Don't think about money or fame; just live a life that suits your own taste
4. Live each day as it comes, cheerfully and without worrying
5. Resist all evils in the world and live a pure and just life
6. Never think of yourself, give everything in service of society
7. Other (Specify)
8. D.K

Figure 2.4 presents the response data. As stated by [13], there has been no significant change in lifestyle preferences since 1973. People's values in the postwar period (e.g., choices 1, 2, 5, or 6) look to have totally lost and most of people prefer choices 3 or 4. Response patterns seem to reflect more or less the mood of home, school, and workplace. Of course, this may not necessarily mean that the majority of people live a cheerful life without worry. Many people may want a lifestyle such as choices 3 or 4 because they can be very stressed in their daily lives and work.

The data seems to coincide with the world trend (cf. International Survey of Youth Attitude by Cabinet Office: https://www8.cao.go.jp/youth/english/survey/2018/pdf_index.html). However, it is important to note that different countries (or individuals)

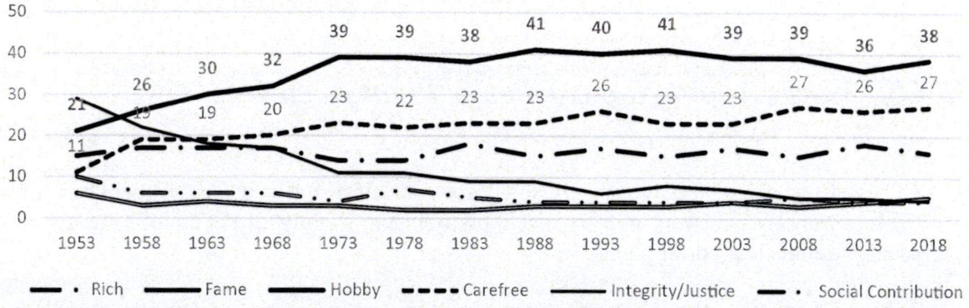

Fig. 2.4 Life Style Preference (#2.4 Attitudes towards Life)

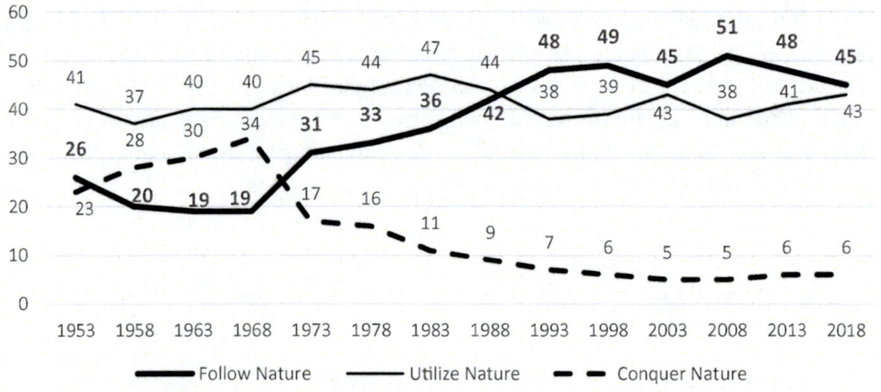

Fig. 2.5 Man & Nature (#2.5)

may give the same answer for different reasons, or different answers for the same reason. Cross-national studies need to investigate the overall situation of each country.

2.4.4 *Humans and Nature*

The following items on humans and nature also show clear changes until 1973, but little has changed since then. The question is #2.5Man and Nature. Figure 2.5 gives the data. See Sect. 4.1.5 of Chap. 4 for a discussion on cross-national data.

#2.5 Man and nature.

(Hand card) Here are three opinions about man and nature. Which one of these do you think is closest to the truth?

1 In order to be happy, man must follow nature
2 In order to be happy, man must make use of nature
3 In order to be happy, man must conquer nature
4 Other(Specify)
5 D.K.

2.4.5 *Worry About Nuclear Power Plant Accidents*

#3.50g. To what extent do you worry, either for yourself or for your family, about each of the following? Would you say very much, pretty much, slightly, or not at all? How about accidents at nuclear power plants?

Table 2.5 shows the changes over time in response to the above item from 1983 to 2018 (remember the disasters of the three-mile island in 1979, Chernobyl

Table 2.5 Worry about on Nucler Power Plants (#3.50g)

Survey Year	Very much	Pretty much	Slightly	Not at all	Other	D.K
2018	27	24	36	12	–	1
2013	**37**	**28**	25	9	–	1
2008	23	22	38	15	–	2
2003	26	23	33	15	0	2
1998	26	26	32	14	–	2
1993	–	–	–	–	–	–
1988	–	–	–	–	–	–
1983	26	20	36	13	1	4

in 1986, and Fukushima in 2011). From 1983 to 2008, about 40–50% of people were "very worried" or "pretty worried" about the nuclear accident. After the 2011 Fukushima nuclear power plant disaster caused by the Great East Japan Earthquake, the percentage inevitably jumped to 65% in the 2013 survey. However, in the 2018 survey, seven years after the catastrophe, the rate returned to around 50%.

Our past studies often reaffirm short-term changes and long-term stability of people's emotions and consciousness [1, 22, 23]. More exactly, on the one hand, people's happiness and unhappiness, as well as their satisfaction and dissatisfaction, can change significantly in the short term (e.g., a day), depending on the objective situation. On the other hand, in the long run, it depends more on personality and national character than on objective situations, and shows a sense of stability such as optimistic or pessimistic tendencies. Short-term fluctuations and long-term stability of people's feelings seem more or less to be reaffirmed also by data on people's concerns about the nuclear accident. (See Note 2 for the universal response pattern of happiness or well-beingness of Japanese people).

2.4.6 Self-Identified Social Stratums

It has been a long time since the issue of disparity (the gap between the rich and the poor) began to be discussed. In the 1980s, it was widely reported that 90% of Japanese consider themselves middle or upper stratum and it was taken for an indicator of the improving standard of living in Japan. This trend was not only in Japan but in many other countries as the World Values Survey (WVS) reported in 1980. Discussions on the expansion of the "middle class" and "educational disparity" were enthusiastically reported in both mass media and academic societies. However, subsequent analysis of time-series data pointed out that the disparity between educational backgrounds in wages was the smallest at that time [24]. This is a warning about discrepancies between the image formed through mass media and the actual situation.

Sakamoto [25] analyzed dozens of variables of data from the national survey of Social Stratification and Social Mobility 1975 (SSM) and clarified that Japanese social stratum identification is subjective rather than objective. In other words, Japanese people distinguish between objective social class (social and economic status, income, etc.) and subjective social stratum (self-evaluation on life one's living conditions or life satisfaction, happiness, etc.). This feature was not seen in the other countries of the 1980 WVS covering 13 countries (Japan, South Korea, Philippine, Singapore, India, Australia, Brazil, United States, UK, Canada, France, Italy, and West Germany).

The development of the middle class is important for democratization. Some scholars believed that economic development would lead to the expansion of middle-class citizens and ultimately advance democracy. But now we are witnessing a counterexample of China. Although China has rapidly developed its economic power and increased the number of rich people, it still maintains an undemocratic system. Economic development may be one of the conditions necessary for successful democracy, but it is not sufficient.

Tables 2.6(a), (b) show the distributions of responses of the following question on subjective social class. The data are from JNCS and from GSS of the United States.

JNCS #1.8 Self-identified Social Stratum

(Hand card) Using the classifications on this card, how would you classify your current standard of living?

1 Upper 2 Upper middle 3 Middle 4 Lower middle 5 Lower

6 Other(Specify) 7 D.K

(GSS) If you were asked to use one of four names for your social class, which would you say you belong in: the lower class, the workingclass, the middle class, or the upper class?

1. Lower class, 2.Working class, 3.Middle class, 4.Upper Class

(There are many complex issues involved in expressing the choices in this question, such as using "working class" instead of "lower class" or "social strata" instead of "social class.")

In general, the results of JNCS are stable over the years. This may be partly because the Japanese tend to avoid extreme answers (see Chap. 3). In general, American people's consciousness often seems to be very sensitive to economic fluctuations, and sometimes significant changes are seen [26]. However, although both Japan and the United States have shown considerable economic ups and downs over the past few decades, Americans also appear to be generally stable with respect to self-identified social classes.

Let's guess the possible interpretations. One may be related to the accuracy of the data. For example, even if we statistically randomly sample about 10,000 of 100 million Japanese voters, the sampling error is estimated to be at least 1% (see Note 1 in Chap. 1). Therefore, a 1% change in all 100 million voters (i.e., 1 million) is likely to be buried in statistical error. However, in the real world, even if the number of unemployed people or homeless people increases by one million, it is a big problem that cannot be ignored. And, whether or not suicide exceeds 30,000 is a big issue in

Table 2.6 (a) Self-identified social stratum in Japanese. (b) Self-identified social stratum in Americans. (c) Self-Identified Social Stratum (APVS [The Asia-Pacific Vales Survey])

a

Survey Year	Upper	Upper Middle	Middle	Lower Middle	Lower
1983	2	12	53	26	5
1988	1	12	52	27	5
1993	1	12	58	23	3
1998	1	10	57	26	4
2003	1	10	57	25	4
2008	2	10	55	28	4
2013	1	13	56	24	4
2018	2	13	54	25	5

b

Survey year	Upper class	Middle class	Working class	Lower class
1972	2	44	47	7
1973	3	46	48	4
1974	3	46	47	4
1975	3	44	48	5
1976	2	48	46	4
1977	4	43	49	4
1978	2	45	47	5

(continued)

Table 2.6 (continued)

b

Survey year	Upper class	Middle class	Working class	Lower class
1980	3	45	46	5
1982	3	41	50	6
1983	4	44	47	6
1984	3	46	46	5
1985	4	47	45	4
1986	3	47	43	7
1987	4	44	46	6
1988	2	48	45	5
1989	4	48	43	5
1990	3	47	46	4
1991	2	49	43	5
1993	3	45	45	7
1994	3	46	46	5
1996	4	45	45	6
1998	4	46	45	5
2000	4	45	46	5
2002	3	46	45	6
2004	3	47	43	7
2006	3	45	46	6
2008	4	43	45	9

(continued)

Table 2.6 (continued)

b

Survey year	Upper class	Middle class	Working class	Lower class
2010	3	42	47	9
2012	4	43	44	8
2014	3	42	47	8
2016	3	41	47	9
2018	3	44	44	8

c

	Survey Year	2010	2010	2011	2011	2011	2011	2012	2012	2012	2013	2013
Item	Category	Japan	USA	Beijing	Shanghai	Taiwan	Hong Kong	South Korea	Australia	Singapore	India	Viet-nam
Social stratum	(1) Upper	0.1	1.5	2.5	1.5	0.9	1.3	0.4	1.6	1.0	10.1	1.7
	(2) Upper Middle	10.9	12.5	11.3	15.6	15.8	9.1	6.2	18.9	9.9	11.6	19.1
	(3) Middle	57.4	52.3	50.9	51.2	59.4	45.6	47.7	58.6	68.5	56.8	70.6
	(4) Lower Middle	24.9	23.6	26.4	22.2	18.6	32.2	36.5	17.0	17.7	12.6	6.4
	(5) Lower	5.9	9.5	8.4	7.7	3.5	10.4	9.0	3.4	2.8	8.7	2.2
	(6) Other	–	0.2	–	0.2	0.2	–	–	0.2	–	0.0	–
	(7) DK	0.8	0.5	0.5	1.6	1.6	1.4	0.3	0.4	0.1	0.1	–

Source the General Social Survey (GSS), NORC. https://gssdataexplorer.norc.org/variables/568/vshow

Japan every year. But that number is two orders of magnitude below the sampling error. In other words, sampling methods such as nationwide surveys are effective for investigating the impact of social problems on the Japanese as a whole, but it is not possible to directly grasp changes in the situation of about one million or less people. From that perspective, considering the Japanese data in Table 2.6(a), it is not always easy to clearly identify the actual changes between adjacent categories beyond the sampling error. However, you may recognize small fluctuations (especially categories between 2 and 3 or between 3 and 4) that may be related to social changes such as the economy.

The second possibility is that the number of people going up and down at the individual level is about the same, and it looks like there is no change overall. To verify that this is happening, it is necessary to follow up with the same respondents over the years. Unfortunately, our style of longitudinal survey does not allow us to do that. I touched on an experimental panel study on individual-level variability and group-level stability [27, preface iii]. (See also Note 7 of Chap. 3).

The third possibility, as a more substantive phenomenon, is that people identify themselves in comparison with their neighbors. For example, Sturgis and Smith [28] report that when asked, "Can most people be trusted?," people often imagine their neighbors and answer this question. Therefore, if people judge the "social stratum" based on the relationship with the neighbors, even if there is an objective change (ups and downs) in the whole society, there may be no change in the overall response distribution (see Note 1).

Table 3 in Yoshino & Osaki [16] shows the distribution of responses for all countries/regions in our cross-country surveys. When choices 2, 3, and 4 are merged as the response category "Middle," about 90% of people, regardless of country or year, choose "Middle." This result may support the third possibility. But I think all the three possibilities are related to the outcome of each country in each situation. For reference, Table 2.6(c) shows the Asia–Pacific Value Survey (APVS) data. In general, the overall pattern is similar in all countries. The political systems of the United States and China are completely different, but there seems to be no remarkable difference between the two. (However, Hong Kong is a little different from Beijing and Shanghai. The proportion of "lower middle" is higher in Hong Kong than in Beijing and Shanghai. Political scientists may find it interesting to further investigate the changes in the economic and political status of the three cities and the public awareness after Hong Kong's return to China in 1997).

2.4.7 *Reward for One's Serious Efforts*

The following items are clearly related to the depression of the Japanese economy over the past 20 or 30 years, including the "lost two decades" (around 1991–2010).

#7.38 Is effort rewarded?

(Hand card) Here are two opinions about striving to achieve one's goals. Which is closest to your opinion?

1. If you work steadily at it, at some point you will be rewarded, without fail
2. No matter how hard you try, there are many times when you won't be rewarded at all

This question was asked before and after the "lost 20 years" (1988 and 2013). When Japan became economically close to the world's top class in 1988, 79% of the respondents said, "If you work steadily at it, at some point you will be rewarded, without fail." And 17% of the respondents said "No matter how hard you try, there are many times when you won't be rewarded at all." However, in 2013 the rates became more pessimistic, at 72% and 26%, respectively. In terms of people's consciousness, basic human relationships are stable over a relatively long period of time, but there are some aspects of consciousness that show close changes as economic and political conditions change. The data presented here is one example of the latter. The other related examples are seen on the item #9.6: Superiority of the Japanese compared to Westerners (Table 2.7) and #9.12: Level of science and technology in Japan (Table 2.8). These examples confirmed that the Japanese lost their self-confidence in many ways during the "lost 20 years." (Note the time delay of effect) .

#9.6 Superior or inferior: Japanese and Western people.

Generally speaking, would you say that the Japanese people are superior to, or inferior to Western peoples?

1 Japanese superior 2 Japanese inferior 3 The same 4 Undecided 5 Other (Specify) 6 D.K. (See Table 2.7.)

9.12 Level of Science and Technology in Japan.

(Hand card) How would you rate the level of science and technology in Japan today? Which of the following four categories would you put it in?

1 Very high 2 Fairly high 3 Fairly low 4 Very low 5 Other (Specify) 6 D.K.

2.4.8 *Japanese People's Fundamental Religious Consciousness and Belief in the Existence of Soul*

Fumi Hayashi [29] found that, regardless of the particular religion, the Japanese have a sort of "soboku-na syukyo-ishiki (fundamental religious consciousness)" that guides their daily lives. Worship of deities in everything has been underlying the Japanese mind as "Shinto" since the early time long before the introduction of Buddhism in the sixth century. Shintoism was dominated by nationalism in the seventh and eighth centuries as a reaction to the newly introduced Buddhism of the time, and was used as a government strategy in the late nineteenth and early twentieth centuries. However, they were far from the original Shinto that has been deeply inside the Japanese mind. Simply put, Shinto is a kind of animism. But the word animism needs attention. Some Western researchers like Taylor ([30], pp.425–426) viewed

Table 2.7 (#9.6) Superior or Inferior: Japanese to Western People

Survey Year	Japanese superior	Japanese inferior	The same	Undecided	Other	D.K
2018	33	10	35	18	0	4
2013	44	7	29	15	0	5
2008	**37**	**9**	**28**	**22**	0	4
2003	**31**	**7**	**31**	**24**	1	6
1998	**33**	**11**	**32**	**19**	0	6
1993	41	6	27	20	0	5
1988	–	–	–	–	–	–
1983	53	8	12	21	2	5
1978	–	–	–	–	–	–
1973	39	9	18	26	0	7
1968	47	11	12	21	1	7
1963	33	14	16	27	1	9
1958	–	–	–	–	–	–
1953	25	20	28	14	21	1

Table 2.8 (#9.12) Level of Science and Technology in Japan

Survey Year	Very high	Fairly high	Fairly low	Very low	Other	D.K
2018	27	60	8	1	0	5
2013	35	52	6	1	0	6
2008	35	51	6	1	0	6
2003	**28**	54	**7**	0	0	10
1998	**24**	57	**10**	1	–	8
1993	46	43	3	0	0	7
1988	43	45	4	1	0	7
1983	–	–	–	–	–	–
1978	37	48	5	1	1	9
1973	33	49	6	2	0	10
1968	–	–	–	–	–	–
1963	–	–	–	–	–	–
1958	–	–	–	–	–	–
1953	–	–	–	–	–	–

animism as a primitive stage of religious development from a Western (or Hegelian) perspective. However, non-Western people do not necessarily think of religion as the axis of one-dimensional development. Instead, as animism represents, they believe that the universal essence of religion beyond words is eternal

JNCS data shows the stable trend that only one-third of Japanese have faith but the majority respect the religious heart/mind over decades. This fundamental religious sentiment is probably deeply rooted to the spiritual religion and ancestor worship that existed before the establishment of today's global monotheism and polytheism. Umehara ([31, p.149]) argues that, like Buddhism or Hinduism, polytheism (worship of nature) and the reincarnation (soul immortality) are at the root of the Japanese mind. And he claims that such religious sentiment was universal throughout the Paleolithic and hunter-gatherer days. He also studies the origin of the Japanese attitude of respecting "donation, self-discipline, humiliation, dedication, Zen, discipline and wisdom" as a basic moral of "Roku Hara Mitsu" (a Buddhism term). He speculates that it was originally in the mind of the Japanese and it has become more refined in the acceptance of Buddhism. Even in regions where world religions such as Buddhism, Catholicism, Protestantism, and Islam dominate, the fundamental religious sentiment of each style may underlie the hearts of the people in each region. If not, how can you explain the non-negligible regional differences between Asia, Europe, and America within the same religion (Buddhist, Christian, or Muslim)?

With regard to this, let us exemplify the following item of JNCS.

#3.5 Life after Death.

"Do you believe in the world after death?"

Unfortunately, this item has not been used continuously, but you can compare the results of the 2nd (1958) and 12th (2008) surveys (Table 2.9). The time difference is half a century.

The percentage of people who denied the existence of a "life after death" was less than half in 2008 compared to 1958. The number of people who believe in "the world after death" has nearly doubled compared to 50 years ago. The gender difference is consistent with each other between 1958 and 2008: the rate of women who believe the world after death is higher than that of men. But the change in the response tendency

Table 2.9 The Rate of those who believe in "the World after Death (Life after Death) (%). (#3.5)

Survey Year	Age (years)	Believe	Believe (Male, Female)	Undecided	Not Believe	Other or DK	Total
1958	20–34	13	(13, 13)	13	66	7	100
	35–49	19	(17, 20)	11	62	8	100
	50–64	33	(23, 42)	10	48	10	100
	65 & other	35	(17, 42)	16	29	20	100
2008	20–	46	(39, 52)	20	30	4	100
	35–49	41	(36, 45)	23	29	7	100
	50–64	36	(30, 42)	22	37	5	100
	65 & over	32	(25, 38)	25	33	9	100

Table 2.10 The Rates of Those Who Believe in God, Life after Death, and Spirit (%) (2010 Japan Survey of the Asia-Pacific Values Survey. See https://www.ism.ac.jp/~yoshino/ap2/jp/table_jp2010.html)

	Gender			Age			Education		
	Total	Male	Female	Young	Middle	Senior	High	Middle	Low
a. Buddha, or other deities such as Jesus, Allah, etc.	81	72	89	81	85	79	80	83	79
b. Life after death	70	62	77	78	82	62	56	70	73
c. A soul (spirit)	81	72	89	89	87	75	76	82	81

on the age group was remarkable. Young people did not believe in the postmortem world more than older people 50 years ago, but now the opposite is true.

There are several possible interpretations for this change. One is that the generation who spent as an infant or adolescent during or shortly after World War II, when society and education changed dramatically, forms a unique cohort with different values than all prewar and postwar generations [5, p. 93, 33]. The people of that generation, who grew up in the postwar democratic era, had relatively rational ideas and were skeptical of their attitude toward religion. The cohort overlaps with the generation of student movement in the 1960s. They were young 50 years ago and are now elderly.

Another interpretation is this. For the last 50 years, Japan had an overwhelming idea of Western style modernization until the 1970s. However, the emergence of environmental problems due to rapid industrial growth may have changed people's hearts in the last 50 years. Unfortunately, this question has not been used consistently for 50 years, limiting its analysis solely from the data in question.

The next item is not from JNCS, but from the Japan 2010 survey of Asia–Pacific Values Survey (APVS).

Q33. Do you think the following items exist? Please answer for each.

a. God, or other deities such as Jesus, Jehovah, Allah, Buddha, etc. b. Life after death c. A soul or a spirit.

(Note: the wordings of these items are translated into each language according to each culture).

(Answer category).

1. Exists/Exists 2. May exist 3.does not exist.

The choices of "1. Exists" and "2. May exist" are classified as "feelings that do not deny existence." For convenience, we call it a "positive answer." But this is not a simple belief in existence, but a kind of complex emotion (i.e., non-rational thinking) that does not deny them completely. Our 2010 Japan Survey (Table 2.10) shows that the majority of the Japanese have positive answers about the existence of "Buddha (or other deities such as Jesus, Allah, etc.)," "Life after Death" and "a soul (spirit)." Let's check the details for each item.

First, 81% of all respondents answered positively on the existence of "a. Buddha or other deities," but there is a gender difference: the rate is 72% for male and 89% for female. On the other hand, there seems to be less difference between age groups. The positive rate is 81% in younger groups under the age of 34, 85% in middle-aged groups aged 35–49, and 79% in older groups aged 50 and older. Also, there is not much difference depending on educational background.

Regarding the existence of "b. Life after Death," 70% of the respondents answered affirmatively, which is slightly lower than "a. Buddha or other deities" mentioned above. By gender, the females' affirmative rate is 77% and the males' rate is 62%, which is somewhat different. By age group, the proportion is about the same (78% of youth, 82% of middle-aged), but 62% of the elderly group, lower than the younger groups. As for the group of 50-year-old and older, 79% of the respondents affirmed that "a. Buddha or other deities" existed, as in other age groups. However, it is interesting that only 62% of the group affirm the existence of "b. World after death." These two items are believed to be related to religious sentiment. It may be necessary to study further the data that some older Japanese do not deny the existence of "Buddha or other deities" but deny the existence of "Life after Death." On the other hand, according to the breakdown of educational background, the proportion of positive answers of high, middle, and low is 73%, 70%, and 56%, respectively. The higher the educational background, the higher the positive rate.

As for the Japanese religious consciousness (nature worship and reincarnation [soul immortality]), this fundamentally animistic or polytheistic attitude often leads to a variety of intermediate attitudes in everyday behavior, with no explicit pros or cons. For the people of modern monotheistic countries, this may be one of the reasons why the Japanese look vague in many ways. This issue will be revisited later in cross-national data analysis [32].

2.4.9 "Obake Chosa" (Ghost Survey): "Rational" and "Non-rational" Types

The research described in the previous section is closely related to the research called "Obake Chosa (Ghost Survey)" that Chikio Hayashi, Hiroshi Akuto, and their colleagues carried out from the 1970s. In that study, they applied multidimensional scaling (MDS) to elucidate the "basic structure of Japanese consciousness" [34, 35]. At first glance, it may sound silly from the research nickname, but it's a deep academic study of survey technology and psychology.

Behind this research was Chikio Hayashi's thoughts. He has taken the lead in the development of postwar scientific opinion polls and various sampling surveys in Japan. However, because of the people's "tatemae" (a kind of social desirability), he couldn't help but feel about the methodological limitations of asking only pros or cons in usual surveys. Of course, "tatemae" is not necessarily worthless. Politics is often driven by a public opinion containing "tatemae." The results of public opinion polls are important basic information for democratic politics. And the results of polls may lead to important policies such as the start or prevention of war. However, in order to deal with complicated social issues such as medical care, environment, and international relations, it is necessary to study the deep structure of people's views of life, nature, and life and death, etc. Therefore, with the intent of trying to get a glimpse of the deep structure of people's minds, they conducted a survey that was significantly different from the usual public opinion survey. The survey covers the items on people's beliefs and interests in ghosts, demons, UFOs, and afterlife.

The basics of public opinion surveys are to analyze the characteristics of the distribution of public opinion in a certain group. Many survey results show that opinion distribution varies according to demographic attributes such as gender, age, educational level, occupation, religion, and classification of residential area such as urban and rural. Usually, beyond these demographic variables, it is difficult to find a particular major group defined by non-demographic variables and showing differences in distribution between groups and other groups. However, in the "Ghost Survey," they investigated the respondent's interests and feelings toward the aforementioned supernatural things, and they have succeeded in identifying the personality classification of "rational people" and "non-rational people." The results of this study were applied to a medical study on the effects of personality in the treatment of cancer [36], a community survey on nuclear power plants [37], etc.

Hayashi and his colleagues started the ghost survey around 1975 and asked about 12 items such as supernatural phenomena (UFOs and Yeti), spiritual phenomena (ghosts and human souls), and fictional creatures (kappa and dragons). Respondents choose from eight choices for each item ("boring," "exist," "nonsense," "scary," "not scary," "interesting/amusing," etc.). The original Japanese question was translated into English for our 2010 US survey as follows.

> Q26. We now would like to ask you some questions about topics that at least some people take seriously. How do you feel about things like "Supernatural Power" and "UFO (Unidentified Flying Object)" on this card? Looking at categories 1 to 8 carefully, please choose the *one* that comes closest to your feeling for each of the following items. [Show Card of items & choices]
>
> a. Supernatural power or psycho-kinesis
> b. UFO (Unidentified Flying Object) or alien
> c. Goblin or ogreGoblin or ogre
> d. Ghost or apparition, curse
>
> (Choices)
>
> 1. Boring
> 2. Would like To Be
> 3. Exist
> 4. Dreadful, Scary
> 5. Would Not Like To Be
> 6. Interesting, Amusing
> 7. Nonsense/Does Not Exist
> 8. Not Dreadful, Not Scary

You find that some dimensions of emotions are mixed in the choices. In a typical survey, the question asks the respondent to choose "yes" or "no" in a particular single dimension. Otherwise, the respondents will be very confused, wondering which dimension to answer (i.e., double-barrel or multibarrel questions). In that sense, the style of choices of the Ghost Survey may look very awkward and misleading. But that is the advanced technology that Hayashi and his team intended. They aimed to identify the dimension that a particular respondent was focusing on from a set of mixed dimension choices, such as emotional or logical.

The results of the data analysis showed that the most common answer for all subjects was "7. Nonsense/Does Not Exist," and very few people affirmed the existence. On the other hand, among the responses showing interest and expectation, the response rate was relatively high especially in the modern ones like UFO. Such interests and expectations tended to be higher in Tokyo than in rural areas, and in younger age groups and in more educated people (although there was a correlation with age and educational level), at least, in the 1970s–1980s. Senior people showed concerns with more traditional things like a ghost. For more detailed results, see Hayashi and his colleagues [26, 29, 34, 35, 38].

Hayashi and his team investigated the relationship between these choices with an additional set of response data on people's attitudes, principles, and values, and found that these choices form two dimensions (inherent dimensions) of "rational type" and "non-national type" [36].

Based on the result, the case where the respondent chose the choice "1" or "7" was defined as "rational type." On the other hand, when any one of the five choices "2," "3," "4," "5," and "6" is selected, it is treated as a "non-rational type." (Because there are few respondents to choose "8," they decided to exclude it in data analysis).

The main points of a series of personality research by Hayashi and his colleagues are as follows.

1. People can be divided into four groups (approximately the same proportion): "rational people," "somewhat rational people," "somewhat non- rational people," and "non-rational people."
2. A "rational person" strongly insists his/her own opinion. Even when persuaded by others (propaganda), there is a tendency not to change one's own opinion. In some cases, he/she shows an extreme change in opinion, which is the exact opposite of what was insisted up to that point (Park and Yoshino [39]).
3. Intermediate types of "somewhat rational" and "somewhat non-rational" are relatively mild and do not strongly insist on their own opinions, but are affected by media reports and others. Often, they tend to be persuaded by other people (propaganda).

(Here, "rational" does not necessarily mean developing perfect logic, but in a word, refers to a personality that strongly insists its own logic and opinion. If the "reasonable" type is too extreme, psychologists could associate it with certain psychological syndromes such as autism spectrum disorders, developmental disorders, or communication disorders).

As mentioned before, this classification has been applied to practical studies such as a medical study on the effects of personality in the treatment of cancer and other diseases [36], a community survey on nuclear power plants [37], and a study on the behavior of voters in the national election.

For example, research on cancer notification shows the following issues. Many of the ordinary people ("non-rational" people) might answer "I want a doctor to notify me if I have cancer" because of a kind of social desirability. But in many cases, their real opinions seem "don't say if it's too serious to cure," or "just vaguely imply, not explicitly." On the other hand, many "rational" people insist that a doctor should "notify," but there are still a few "rational" people who do not want to be notified. And, importantly, if the last type of person is notified when they are actually suffering from cancer, it can adversely affect their medical condition [36]. (Note. Due to significant advances in cancer care over the last few decades, Hayashi's research may not always be applicable to the recent cancer notification problem. But it may still be applicable to certain serious illnesses).

As for the study of nuclear power plants, Hayashi and Morikawa [37] described people's attitudes and the way of their opinion changes, as follows. Both those who strongly support nuclear power plants and those who strongly oppose it have a high level of scientific knowledge. Also, they are hard to be persuaded by others, and hard to be swept away by propaganda. But sometimes, for their own reasons, and no other persuasion, they may suddenly change their opinions to the exact opposite extreme. These people correspond to the "reasonable" type. On the other hand, although the general public of the "non-rational" type does not insist, they tend to change their opinions depending on how the media report a nuclear accident.

These studies also reaffirm that when political militants change their attitudes, they change their attitudes to the extreme. Therefore, the political scaling from left to right is not one-dimensional, but two-dimensional, classifying the "rational" and "non-rational" groups in the first dimension and the left and right of each group in the second dimension. (cf. Sect. 3.2.3 of Chap. 3 of this book).

I will not further develop this analysis here. However, I think this is important in studying various styles of conflicts between militants and civilians, domestic and international. (For cross-national study of the Ghost Survey, see Park and Yoshino [39]).

This chapter has reviewed the data of the Japanese National Character Survey. For more advanced research, we need to proceed to the analysis of cross-national comparative survey data.

Notes

Note 1. According to the Asahi Shimbun [40], regarding the positioning of political attitudes (conservative and liberal analysis), the distribution pattern did not change significantly even if the respondent groups were divided into several groups. Each one seems to identify his/her political attitude in comparison to others. The general characteristic of people's self-identification regarding social issues (relativeness) seems to be reflected also in the distribution of political attitudes.

Note 2. Many surveys of Japanese well-being, happiness, or life satisfaction show bimodal response distributions on scales such as the 10-point scale, with peaks around 5 and 8. This can be interpreted as having two major groups: One group to choose the median (i.e., "5" on a 10-point scale) and another group to choose the median on the positive side of the scale (i.e., "7" or "8"). (See https://www.esri.go.jp/jp/prj/current_research/shakai_shihyo/pdf/panel_dis4.pdf and https://www5.cao.go.jp/keizai2/manzoku/pdf/report01.pdf, Table 17). From these observations, we warn against using such a 10-point scale to see long-term changes or to see the effect of a particular policy (Yoshino, [41]). We also need to be careful about comparisons between countries on such a point scale (cf. OECD) [42]. Such a point scale may be a measure of a stable and general response trend of an individual or country.

2.5 Summary

The Japanese National Character Survey (JNCS) shows longitudinal change of Japanese attitudes, opinions and values on various topics in daily life over 65 years since 1958. Above all, the gender difference of change of responses to the question "If you were born again, would you like to be born as a boy or a girl?" symbolizes the longitudinal change of postwar Japan regarding not only on the issue of gender equality but social attitudes and values.

The other earlier main findings up to the middle of 1980s can be summarized as follows:

(1) Since the beginning of the Meiji era (1868), Japanese people have maintained an emotional conflict between Japanese tradition and modernization (Westernization) even after World War II. Some Japanese traditional attitudes such as "giri-ninjyo" attitude, religiosity, and basic ways of thinking remain unchanged. Therefore, in the process of accepting Western things with the Japanese spirit and emotions, it seemed that both the modernization of Japanese culture and the Japanization of Western culture were mixed.

(2) Regarding religion, Japanese tend to have religious faiths as they grow older. On the other hand, the majority of Japanese consider religious heart/mind to be important, even though only a third of them have religious beliefs. In general, this trend has been stable, but has become somewhat ambiguous due to global secularization and due to the cohort effect of the age group born between 1935 and 1945. This age group overlaps with the generation of the student movement in the 1960s. They were born during World War II and experienced a rapid change from war to postwar. Inevitably they became skeptical of authorities, including politics and religion, and grew more "rational" in the age of postwar democracy. Their attitudes and values are very different from both older and younger generations.

(3) The period from 1973 to 1978 was a significant turning point in the postwar Japanese attitude. The 6th survey (1978) showed signs of the birth of a new system of beliefs, indicating a significant turning. These signs appeared in the young group (aged 20–24) born about 10 years after World War II. The young group were born at the beginning of industrial and economic development but witnessed that, around 1970, Japan faced pollution problems as a result of its rapid industrialization. In the 1980s when Japan ranked near the top of the world economy, those young people were named "new mankind" because the elderly could hardly understand their attitudes and values.

(4) In the 1980s, Japan was economically successful, but the Soviet Union was on the verge of the collapse and the United States was suffering from domestic and trade deficits, so Japan lost the ideal model for further development and must create its own model. The government repeated substantial economic and political reforms under the name of "globalization," but ultimately resulted in the "lost 20 years" of the 1990s and the 2000s. Many items of JNCS showed a break in change or a reversal in the direction of change during 1973–1978.

> Since 1978, the response distribution for many items fluctuated up and down. In 2011, the Great East Japan Earthquake and the subsequent Fukushima nuclear disaster had a significantly strong direct impact on the Japanese. In subsequent developments of research, however, we gradually came to learn that human or interpersonal relationships have been stable in the deep structure of people's consciousness for decades in spite of superficial change of styles.

Inglehart [11] used the JNCS data and recognized an important issue of comparison between East and West nations: the items in the list of postmaterialist that were valid in the West were neutral or even reversed in Japan. Therefore, scaling, or more generally research methods, should be considered in a cultural context. Western researchers often use analytical hypothesis testing and theory-building, while Eastern researchers work on a holistic data-driven approach.

Finally, the "Obake Chosa" (Ghost Survey) identified the difference in personality between the "rational type" and the "non-rational" type. Here, "rational" does not mean developing perfect logic, but refers to a personality that strongly insists its own logic and opinion. "Rational" people tend not to change their opinions when persuaded by others. But in some cases, they suddenly change their opinion to the extreme opposite of what was previously insisted. The intermediate type between "rational" and "non-rational" is relatively mild and does not insist one's opinion, but is influenced by media reports or propaganda. These differences are important for understanding the deeper structure of survey responses, both at the individual and national levels.

References

1. The Japanese National Character Survey Committee. (1961). *Nihon-jin no Kokuminn-sei* (Japanese National Character). Tokyo: Shisei-do.
2. The Japanese National Character Survey Committee. (1970). *Dai-2 Nihon-jin no Kokuminn-sei* (Japanese National Character, Vol. 2). Tokyo: Shisei-do.
3. The Japanese National Character Survey Committee. (1975). *Dai-3 Nihon-jin no Kokuminn-sei* (Japanese National Character, Vol. 3). Tokyo: Shisei-do.
4. The Japanese National Character Survey Committee. (1982). *Dai-4 Nihon-jin no Kokuminn-sei* (Japanese National Character, Vol. 4). Tokyo: Idemitsu-shoten.
5. The Japanese National Character Survey Committee. (1992). *Dai-5 Nihon-jin no Kokuminn-sei* (Japanese National Character, Vol. 5). Tokyo: Idemitsu-shoten.
6. Hayashi, C., & Sakuraba, M. (2002). *Nihon-jin no senzairyoku (Japanese potential).* Tokyo: Kodansha.
7. Hayashi, C., & Suzuki, T. (1984). Changes in belief system, quality of life issues and social conditions over 25 years in post-war Japan. *Annals of the Institute of Statistical Mathematics, 36*(1), 135–161.
8. Benzecri, J-P. (2019). *Correspondence analysis handbook*. Routledge.
9. Nishisato, S. (1994). *Elements of dual scaling: an introduction to practical data analysis.* Psychology Press.
10. Young, F. W., de Leeuw, J., & Takane, Y. (1976). Regression with qualitative and quantitative variables: An alternating least squares method with optimal scaling features. *Psychometrika, 41,* 505–529.

11. Inglehart, R. D. (1982). Changing values in Japan and the west. *Comparative Political Studies, January.* 445–479.
12. Flanagan, S. C. (1979). Value Change and partisan change in Japan: The silent revolution revisited. *Comparative Politics, 11,* 253–278.
13. Sakamoto, Y. (2005). A statistical research on the Japanese national character: General trends in attitude over the past half fifty years. *Proceedings of the Institute of Statistical Mathematics, 53*(1), 3–33.
14. Inglehart, R. D. (2018). *Cultural evolution: People's motivations are changing, and reshaping the world*. London: Cambridge University Press.
15. Yoshino, R. (2015). Cultural manifold analysis (CULMAN) as a paradigm of cross-national comparative surveys on national character. *Proceedings of the Institute of Statistical Mathematics, 63*(2), 203–228.
16. Yoshino, R., & Osaki, H. (2013). Subjective social class, sense of satisfaction, and sense of trust – a note on psychological scales of social surveys [in Japanese]. *Japanese Journal of Behaviormetrics,* 40, 2, 97–114. https://www.jstage.jst.go.jp/article/jbhmk/40/2/40_97/_article/-char/ja/, Retrieved January 11, 2016.
17. Yoshino, R., Shibai, K., & Nikaido, K.(eds.). (2015). Sense of Trust—Summary Report: The Asia-Pacific Values Survey - Cultural Manifold Analysis(CULMAN) on People's. *ISM Survey Research Report*, No. 117.
18. Meyer, E. (2014). *The culture map: Decoding how people think, lead, and get things done across cultures.* NY: Public Affairs.
19. Sakamoto, Y. (2000). How has the Japanese way of thinking changed? A half century of the statistical survey of the Japanese national character. *Proceedings of the Institute of Statistical Mathematics, 48*(1), 3–32.
20. Nakamura, T. (1986). Bayesian cohort models for general cohort table analyses. *Annals of Institute of Statistical Mathematics, 38,* 353–370.
21. De Leeuw, E., Hox, J., Silber, H., Struminskaya, B., & Vis, C. (2019). Development of an international survey attitude scale: Measurement equivalence, reliability, and predictive validity. *Measurement Instrument for the Social Sciences, 1*(9), 1–10. https://doi.org/10.1186/s42409-019-0012-x.
22. Kahneman, D., & Krueger, A. B. (2006). Developments in the measurement of subjective well-being. *Journal of Economic Perspectives, 20*(1), 3–24.
23. Hofstead, G., Hofstead, G. J., & Minkov, M. (2010). *Culture and organizations—software of the mind* (3rd ed.). NY: McGraw-Hill.
24. Hashimoto, K. (2009). *Postwar history of social gap.* Tokyo: Kawade-book.
25. Sakamoto, Y. (1987). What makes the difference in social stratum identification? *Proceedings of the Institute of Statistical Mathematics, 35*(2), 233–255.
26. Yoshino, R., Hayashi, F., & Yamaoka, K. (2010). *Kokusai- hikaku deta no kaiseki (Analysis of comparative survey data).* Tokyo: Asakura-shoten.
27. Yoshino, R. (2001). *Kokoro wo hakaru (Measuring the mind).* Tokyo: Asakura-syoten.
28. Sturgis, P., & Smith, P. (2010). Assessing the validity of generalized trust questions: What kind of trust are we measuring? *American Sociological Review, 21*(6), 690–695.
29. Hayashi, F.(2006). A study of religious faith and religious feeling. *Japanese Journal of Behaviormetrics*, 33(1),3–24.
30. Taylor, E. B. (1903[1871]). *Primitive Culture. 2 vols.* London: John Murray & Co.
31. Umehara, T. (2002). *Creation of universal human philosophy* (in Japanese). Umehara Takeshi Collection 17. Tokyo: Shogaku-kan.
32. Yoshino, R. (1992). Superculture as a frame of reference for cross-national comparison of national characters. *Behaviormetrika, 19*(1), 23–41.
33. Yoshino, R. (2013). Trust of nations on cultural manifold analysis (CULMAN): Sense of trust in our longitudinal and cross-national surveys of national character. In N. I. Dryakhlov, A. Ishikawa, A.B. Kupreychenko, M. Sasaki, Z. T. Toshchenko, & V. D. Shadrikov (eds.), *Trust in society, business and organization: Proceedings of the conference, "business, society, human"* (pp. 213–250). Moscow: National Research University.

34. Hayashi, C. (Ed.). (1984). *Practical application of multidimensional analysis (in Japanese).* Tokyo: Science-sha.
35. Hayashi, C., & Akuto, H. (1976). *Multidimensional analysis.* Tokyo: Science-sha.
36. Hayashi, C. (1996). ***The Japanese mind and cancer notification*** (in Japanese). Japan Cancer Disease Treatment Study Group QOL Group.
37. Hayashi, C., & Morikawa, S. (1994). National character and communication: How to deal with people's attitude towards the nuclear power generation (in Japnese). *INSS JOURNAL, 1,* 93–135.
38. Hayashi, F. (2010). Contemporary Japanese religious mind—based on survey data from the Japanese national character and other cross-national Surveys. *Proceedings of the Institute of Statistical Mathematics, 58*(1), 39–59.
39. Park, Y., & Yoshino, R. (2015). "Obake (Ghost) surveys" revealing underlying structure of heart and mind: Some relevant data from Asia Pacific Values Survey (APVS). *Proceedings of the Institute of Statistical Mathematics, 63*(1), 163–195.
40. Ishihara, Y. (2010). Liquefied political party support structure. *Journalism*, 240 (May 2010), 56–69.
41. Yoshino, R. (2014a). Kouhuku-do ha seisaku-kagaku no tameni sokutei kanou ka? [Can "happiness" be measured as scientific scale for evidence-based policymaking? *Keikaku-gyosei* (Planning and Public Management), 37, 2, 35-40.
42. OECD (Organization for Economic Cooperation and Development). (2013). OECD guidelines on measuring subjective wellbeing. https://www.oecd.org/statistics/oecd-guidelines-on-measuring-subjective-well-being-9789264191655-en.htm.

Chapter 3
Cross-National Comparative Survey on People's Consciousness

"Make your facts comparable before deducing causes." Florence Nightingale [1]

3.1 History and Tasks of Our Cross-National Survey

3.1.1 New Development of JNCS

Beginning in 1971, the Japanese national character survey was expanded to cover cross-national surveys for acquiring a more advanced understanding of the Japanese national character in a comparative context [2]. In designing these cross-national survey questionnaires, we selected items to compare people's social values, their ways of thinking, their emotions, etc. More precisely, their cultural identities, interpersonal relationships, leadership qualities, religious attitudes, and social values related to science and technology, politics, economy, social security, freedom of speech, etc. These aspects may yield information on psychological distances between nations or races, as evidenced by certain response patterns. Based on the survey data, our research team has been attempting to quantitatively ascertain peoples' attitudes and values, as well as the characteristics of their civilizations. Our ultimate goal is to develop a *behaviormetric* study of civilizations based on social survey data as this will provide fundamental information for the peaceful development of the world (Hayashi [57]).

In the study of comparing various countries, the major issue is how cross-national comparisons are possible, under problems related to translation and differences in survey methods unique to each country. We are studying the methodology for pursuing this *cross-national comparability*, not simply comparing and interpreting survey data superficially. In this pursuit, we are trying to build a statistical philosophy called the "Science of Data " (see Preface of this book and [58]) for the quantitative study of civilization ([57, 3, 4, 16, 24, 26]).

R. Yoshino, *Cultural Manifold Analysis on National Character*,
Behaviormetrics: Quantitative Approaches to Human Behavior 10,
https://doi.org/10.1007/978-981-16-1673-0_3

Even with funding, statistically rigorous social survey cannot be conducted in all countries. In some countries, accurate and detailed census statistics and an accurate list of citizens or voters may not be available in surveys. In non-democratic countries, social survey is effectively limited, especially for sensitive political or religious issues.

In Japan, an almost complete list of voters and residents is available for polls, allowing a relatively ideal statistical random sampling. However, for cross-national surveys, it may be necessary to use a method that is significantly different from that used in Japan, depending on the circumstances of each country. Questionnaire translation is another major issue when comparing countries with different cultures and languages.

As a scientific study, cross-national survey must overcome these multifaceted methodological issues. Therefore, an important issue in our study is to identify the conditions under which a cross-national comparison of survey data is warranted, or the degree of a cross-national comparability. This problem entails several analytical and methodological sub-problems. These relate to: (1) translation (i.e., the questionnaire items must be written to keep the same meanings in different languages), (2) comparison of data sets collected using different sampling procedures in different countries, (3) characterization of nations or peoples (i.e., ascertaining the aspects and degrees of similarity and dissimilarity of certain nations or peoples), (4) description of nations or peoples in terms of a common logic (or framework of thinking) rather than a logic particular to a certain nation or people, etc. (Note 1).

These sub-problems are complementary in the sense that their solutions are interrelated. Some of our findings on these topics have been reported in many publications (See [4, 5, 14, 24, 25, 35, 45, 51] and their references). These studies have led us to frequent utilization of multidimensional data analysis called Hayashi's Quantification Method III [6, 46], to attempt to overcome the problems of translation and the use of different sampling methods. Regarding the fourth issue above, we are struggling to establish a methodology for expressing the nation and the people in common framework. We believe that the research paradigm called CLA or CULMAN (Sect. 3.2) provides a framework for developing empirical social sciences to understand the rise and fall of civilization and promote mutual understanding between different cultures. We also hope that this paradigm will help us predict collective action and international political commitment in certain countries in order to avoid conflicts and develop world peace.

For the purpose to deepen our understanding of Japanese national character, it seemed better to start comparing the Japanese in the homeland with Japanese immigrants abroad. Thus a full-fledged overseas survey by the Institute of Statistical Mathematics (ISM) started as Hawaii residents survey in 1971. Many Hawaiians are of Japanese descent ("Nikkei"). This is closely related to the development of Cultural Link Analysis, which will be explained in Sect. 3.2 [7, 14, 52]. Table 3.1 shows our major cross-national surveys.

The rest of this section discusses some basic cross-national comparison (translation, sampling, etc.) issues and the knowledge that has accumulated over the decades. Section 3.2 presents research paradigms for cross-national comparison. The basic

Table 3.1 Longitudinal and cross-national surveys by ISM

· **1953—present (every 5 years) The Japanese National Character Survey (JNCS)**

· **Surveys on Japanese Americans of Hawaii & the West Coast, and Japanese Brazilians**

1971: Honolulu Residents with Japanese ancestry (Chikio Hayashi, Hirojiro Aoyama, Shigeki Nishihira, Tastuzo Suzuki [ISM] Kikuo Nomoto [National Institute of Japanese Language]; Yasumasa Kuroda [Hawaii University])

1978: Honolulu Residents, Americans in the Mainland; 1983: Honolulu Residents

1988: Honolulu Residents; 1991: Japanese Brazilians (JB) in Brazil (Chikio Hayashi, Hiroshi Midzuno, Tatsuzo Suzuki, Fumi) Hayashi, Ryozo Yoshino [ISM]; Katsuzo Yamamoto,Takeo Kawai, Katsunori Wakisaka, Susumu Miyao, Koich Mori [San Paolo Human Science Institute]

1998: Americans with Japanese ancestry on **the West Coast of USA** (Ryozo Yoshino, Chikio Hayashi, Tatsuzo Suzuki, Fumi Hayashi, Tadahiko Maeda, Kazue Yamaoka, Hiroko Kawahara, Masakatsu Murakami [ISM]; Frank Shotaro Miyamoto Tetsuden Kashima, Tsukasa Namekata [Washington University]; Stefan Fugita [Santa Clara University]

1999: Honolulu Residents with Japanese ancestry (Ryozo Yoshino, Chikio Hayashi, Tatsuzo Suzuki, Fumi Hayashi, Yuejun Zheng, Tadahiko Maeda, Masamichi Sasaki, Kazue Yamaoka, Hiroko Kawahara [ISM]; Kuroda Yasumasa [Hawaii University] Frank Shotaro Miyamoto, Tetsuden Kashima [Washington University]; Stefan Fugita [Santa Clara University])

· **1987–1993 Seven Country Survey (Japan, USA, UK, W. Germany, France, Italy & the Netherlands)**

(Chikio Hayashi, Tatsuzo Suzuki, Fumi Hayashi, Ryozo Yoshino[ISM]; Ichiro Miyake [Kobe University], Masamichi Sasaki [Hyogo-Kyoiku University])

1987: UK, West Germany & France; 1988: U.S.A. & Japan; 1992: Italy; 1993: The Netherlands

· **2002–2005 East Asia Values Survey (EAVS)**: Japan, China [Beijing, Shanghai], Hong Kong, Taiwan, South Korea & Singapore (Ryozo Yoshino, Fumi Hayashi, Kazue Yamaoka, Yuejun Zheng, Tatsuzo Suzuki, Masamichi Sasaki, Msakatsu Murakami Yasumasa Baba, Tsuchiya Takahiro, Tadahiko Maeda, Takahiro Hoshino, Wataru Matsumoto, Hiroko Kawahara [ISM] Keiko Nakao [Tokyo Metropolitan University]; Yanyun Zhao, Jang Qiyan, Yuan Wei [Renmin University of China] ZHOU Guomo [Zhejiang University]; Ben-Chang Shia [Fu Jen Catholic University], Hing Po Lo, Agnes Law [Hong Kong City University]); Seung-Geun Park [**National Statistical Office** in South Korea]

· **2004–2009 Pacific-Rim Values Survey (PRVS)**: Japan, China [Beijing, Shanghai], Hong Kong, Taiwan, South Korea, USA Singapore, Australia & India (Ryozo Yoshino, Fumi Hayashi, Kazue Yamaoka, Tatsuzo Suzuki, Takahiro Hoshino, Yuejun Zheng) Wataru Matsumoto, Yoshihiro Miyoshi, Hiroko Kawahara Masamichi Sasaki, Yasumasa Baba [ISM]; Yanyun Zhao, Jang Qiyan Yuan Wei [Renmin University of China]; ZHOU Guomo [Zhejiang University]; Ben-Chang Shia [Fu Jen Catholic University] Hing Po Lo, Agnes Law [Hong Kong City University]

· **2010–2014 Asia–Pacific Values Survey (APVS)**: Japan, China [Beijing, Shanghai], Hong Kong, Taiwan, South Korea USA, Singapore, Australia, India & Vietnam (Ryozo Yoshino, Fumi Hayashi, Kazue Yamaoka, Tatsuzo Suzuki, Tsuchiya Takahiro Tadahiko Maeda, Yuejun Zheng, Wataru Matsumoto, Park You-Sung, Akira Horoiwa, Taisuke Fujita, Kosuke Nikaido, Hiroko Osaki Kiyohisa Shibai, Hiromasa Hattori, Hiroko Tsunoda, Masamichi Sasaki, Yasumasa Baba [ISM]; Yanyun Zhao, Jang Qiyan, Yuan Wei [Renmin University of China]; ZHOU Guomo [Zhejiang University] Ben-Chang Shia [Fu Jen Catholic University]

Note For details of the member lists and statistical tabulations, see the following websites
https://www.ism.ac.jp/editsec/kenripo/contents.html https://www.ism.ac.jp/~yoshino/index.html

steps for a cross-national survey are outlined in Sect. 3.3. Finally, Sect. 3.4 describes general country-specific response tendencies in attitude and opinion surveys (e.g., Japanese often tend to avoid explicit answers to questions).

3.1.2 Translation and Back-Translation

If you want to include a certain question item that has never been used in a cross-national survey, you need to translate it into another language for a cross-national survey. This section describes the case where the English version of questionnaire was created from the Japanese version using the translation/back-translation methods [8].

The procedure for translation/back-translation is as follows:

(1) A bilingual (preferably a native English speaker) translates Japanese questions into English.
(2) Then another bilingual (preferably a native Japanese speaker) translates the English translated version back into Japanese.
(3) Compare the back-translated Japanese wording with the original Japanese version. If there is no difference between the two, we may assume that the English translated version can be used as the equivalent of the Japanese version (unless there is any evidence against that assumption). You may need to change the translation if there is an inherent difference between them, and the translation/back translation process needs to be repeated if necessary.

However, we came across some cases difficult to determine if particular differences of wordings are ignorable or not. Therefore, the following experimental survey was conducted.

Two homogenous Japanese samples (each 1000 respondents) were created by split-half method, and one of these two samples (Japan A) was given questionnaire A and the other (Japan B) was given questionnaire B. Questionnaire B consisted of the original Japanese question items, while questionnaire A consisted of the back-translated Japanese question items. The wording for some items was exactly the same in both questionnaires A and B, but there were slight differences in the wording for some other items. We compared the response distributions of Japan A and Japan B.

In the following example, there was no difference in the wordings of question, but there were some differences in wording in the response choices. The question was:

> Q33. If you were asked to choose the two most important items listed on the Card, which two would you choose? (Select two)

In the original Japanese, the choices were:

a. 親孝行 (Filial piety) 73.2%
b. 恩返し (Repaying to one's benefactor) 45.8%
c. 個人の権利 (Respect for the rights of the individual) 37.7%

d. 個人の自由(Respect for the freedom of the individual). 36.6%

(Here the English in the parentheses is the tentative word-by-word translation by the present author for the readers' convenience. Percentages following each choice indicate the response percentages.)

These items were translated into English as follows.

a. Filial piety/Love and respect for parents
b. Repaying people who have helped you in the past
c. Respect for the rights of the individual
d. Respect for the freedom of the individual

Then the English was back-translated into Japanese as follows.

a. 親孝行/親に対する愛情と尊敬 (Filial piety/Love and respect for parents) 77.7%
b. 助けてくれた人に感謝し、必要であれば援助する (Express gratitude to people who have helped you in the past and repay them when necessary) 56.8%
c. 個人の権利 (Respect for the rights of the individual) 25.2%
d. 個人の自由 (Respect for the freedom of the individual) 32.8%

(Here again, the English in the parentheses is the tentative word-by-word translation by the present author.)

Thus, in the back-translation, the Japanese expressions of Choice a and Choice b were changed, but the Japanese expressions of Choice c and Choice d were not changed. In the comparative survey, "b" made a significant difference in the response rate (11%). There was no change in the wording of "c," but the largest response change was observed in that item (12.5%), probably due to the impact of the change in "b."

Figure 3.1a shows the response distribution for each corresponding common item for Japan A and Japan B , where the wording of question and choices is totally the same. This response difference can be primarily due to statistical sampling errors. Figure 3.1b shows the distribution of corresponding items, such as the example in Q33, where the wording of the question or choice is slightly different. The variance in Fig. 3.1b is larger than the variance in Fig. 3.1a. Thus, even small differences in wording sometimes can result in 10–20% differences in response rates. Therefore, great care should be taken when comparing response rates from different countries where different languages are used. Even if there are some differences in the response distributions, it may not be immediately possible to conclude whether it is a real difference or the effect of different wording. Some may think that a 10–20% difference in the distribution of responses due to slight differences in wording is too large to guarantee cross-national comparison. However, try applying multidimensional pattern analysis such as Hayashi's Quantification Method III (Hayashi [6] to data from several countries or regions for all items (or a particular set of items) in the same questionnaire. In the output Fig. 3.2, the distance between countries/regions corresponds to the similarity of the response patterns. You can see that the difference between Japan A and Japan B is small enough in the whole figure.

These observations lead to the following conclusions. First, comparing the response distributions of two countries for one item can be difficult to determine if

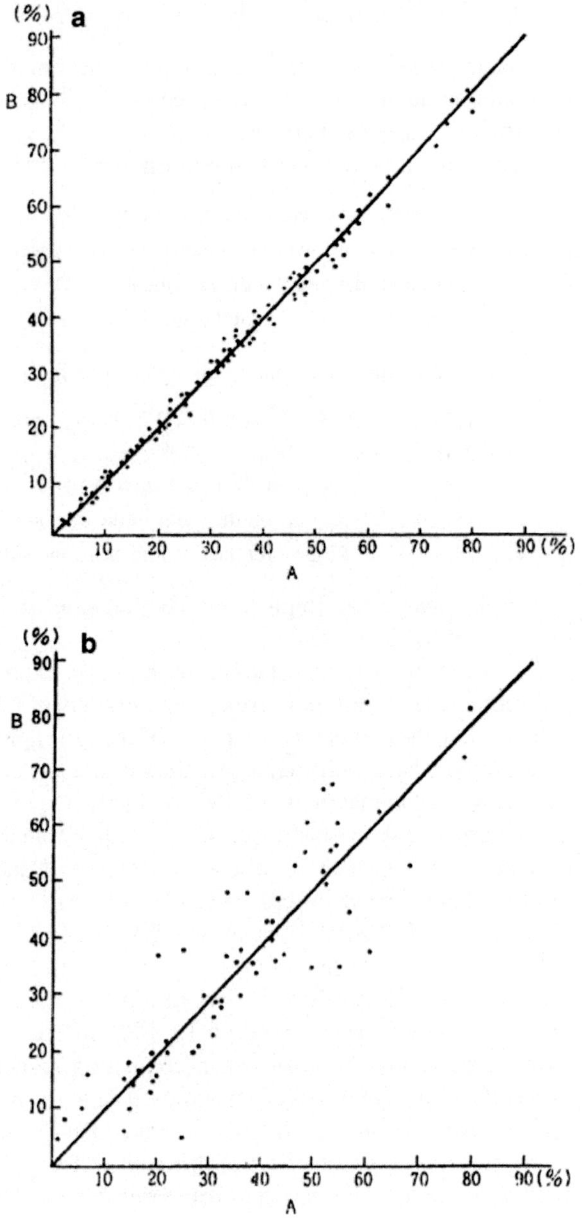

Figure 3.1 **a** Comparison of response percentages of Japan A (back-translated Japanese questionnaire) and Japan B (original Japanese questionnaire). In the case where the wording is completely the same in both Japan A and Japan B, the variance may be comparable to sampling errors. **b** In the case where there are some wording differences between Japan A and Japan B, the differences in response percentages sometimes amount up to 10–20%

the difference is essential or due to translation (unless the difference is large enough, say, more than 20%). However, second, when comparing response patterns from multiple countries over many items, the effects of subtle differences in question and choice expressions may be negligible (Of course, this discussion temporarily sets aside the semantic issues of the cultural context).

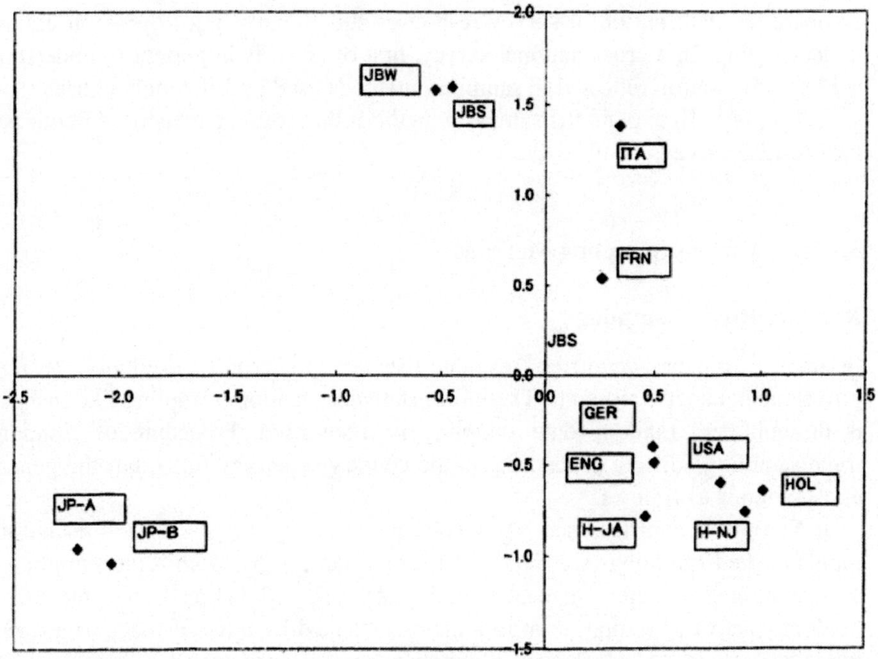

Fig. 3.2 Hayashi's Quantification Method III (QMIII) applied to data of all common items from several countries/areas. The effect of wording differences (Japan A and Japan B) or sample weighting (JBS and JBW) is so small in a global comparison that it may be negligible

3.1.3 Sampling Methods

Over the last 20 or 30 years, the privacy of respondents and the safety of interviewers have led to a significant shift in survey mode from traditional face-to-face surveys to landline telephone, mobile telephone, or online surveys. However, these new survey modes do not always ensure the rigorous statistical randomness and population coverage that is important for statistical inference.

The basic policy of our cross-national comparative survey is to conduct face-to-face interviews using statistically rigorous sampling and to collect reliable data as a representative sample from each country. This method is most reliable to obtain statistically high-quality data. However, it is not always possible to carry out statistical sampling as strictly as in Japan. Therefore, each country has invented its own sampling method.

Given the fact that many important decisions are greatly influenced by opinion polls, it is important to respect and learn from each country's research methods that have developed in the history of each country. Also, based on the experience of cross-national comparative studies over the past 30 years, I believe that the sampling methods commonly used in each country reflect important aspects of the country's political, economic, and social situations. The main purpose of our study is not to compare the distribution of survey responses superficially, but to promote mutual understanding. In a cross-national survey, first of all, it is important to understand the key information reflected in sampling methods used in different countries.

Below, we will explain the sampling methods that were actually used in our past face-to-face surveys (Table 3.1).

3.1.3.1 Various Sampling Methods

Random-Route Sampling

In many European countries, resident lists are not readily available for non-governmental social research. Therefore, statistical random sampling like in Japan is difficult, and random-route sampling is often used. Procedure of "random-route sampling" differs depending on the country or survey firms, but the general explanation is as follows.

In European countries, as in Japan, nationwide surveys usually use multistage and multistratified sampling (see Sect. 1.2.1.2 of Chap. 1). Random-route sampling is used at the stage of selecting respondents in each selected district. If you statistically randomly select a starting point in a district selected by a list of roads or a region map, the interviewer will start visiting houses from door to door, for example, every three doors to the left of the selected road. At each door, he/she asks to interview a particular respondent selected by Kish method (using a random number table), recent birthday rules, etc. The interviewer will continue to visit houses until he/she gets a certain number of fixed valid respondents (e.g., 10 or 20 respondents per district).

Although unlikely to be strictly distinguished, it is called **random-route sampling** in Europe, **random walk** in the United States, and **right hand technique** (walking clockwise along a selected road) in India (It is important to note that technical terms may not match in different countries, as each survey firm accumulates its own know-how and keeps it confidential as an asset. In particular, Japan's postwar survey system has been established independently of other countries, so it is not always easy to match the technical terms of Japan and other countries).

Random-route sampling is not about interviewing a predetermined sample of respondents, but looking for people who will participate in the survey until a given number of respondents are obtained. Random-route sampling does not use the concept of "response rate (contact rate and cooperation rate)." This is not desirable from a statistical point of view because the difference between respondents (participating in the survey) and non-respondents (declining to participate in the survey) can influence the statistical estimation of the total population distribution from the observed sample. However, it is often used in Europe as a method with historically accumulated know-how. (Although the exact "response rate" cannot be calculated, in the 1992 Italian survey and the 1993 Dutch survey of our cross-national research, the cooperation rate was about 30–40%, considering the total number of households visited in the survey.)

Quota Sampling

With quota sampling, the procedure up to selection of sampling districts is the same as in Japan, but when selecting respondents at each selected district, a specific quota table is used. This table should match the census data for pre-specified attributes such as gender, age group, and ethnicity. For the percentages of these attributes under consideration, in many cases, the entire set of quota tables, rather than each quota table for each district, is formatted to match the census data. Then find out the pre-specified number of respondents (e.g., 10 or 20) according to the quota table. For example, one male Asian of 20s, two female white Caucasian of 40s, and so on.

Strict statistical random sampling, such as used in Japan, selects a sample of respondents without biasing all attributes at least probabilistically. With regard to quota sampling, it is not immediately justified to consider the collected data as a representative sample, as the degree of bias due to unspecified attributes (education, income, religion, etc.) cannot be known in advance. Responses to the survey assume that factors other than pre-specified attributes are not very effective, but this assumption may turn out not to be correct. On the other hand, pre-allocation of educational background, income, religion, etc. will significantly increase costs of the survey procedure. In addition, there is no limit to the number of "attributes," and some are difficult to identify technically or ethically, such as intelligence, personality, or health. Also, strictly speaking, it is statistically more accurate to create a quota table, taking into account both the total and regional population of the attributes of interest, but in practice this is very difficult (cf. The 2011 Shanghai Survey involved such a difficult sampling process. Yoshino et al. [9]).

However, quota sampling is often used in many countries where sampling lists are not available for social survey. This sampling can be viewed as a method developed throughout the long history of politics seeking equal rights to vote without discrimination by age, ethnicity, or gender.

Quarter sampling is said to be more appropriate when combined with random-route sampling because it reduces interviewer arbitrariness in selecting respondents. But interviewers' reports on random-route sampling of the 2006 and 2011 surveys in the United States and and the 2012 survey in South Korea found that cooperation rate was too low and interviewers visited almost every home in the selected sampling area. So, random-route sampling was effectively about the same as quota sampling. We shouldn't be satisfied with a theory on desk, but do our best to understand what is really happening in the survey process.

Random-route sampling or quota sampling is not statistically desirable, but they appear to be used often in Europe and the United States for market research. However, market research and opinion polls have different purposes. Market research seeks to identify preferences and buying habits of groups identified by attributes such as gender, age, educational background, income, place of residence, etc. On the other hand, in polls of democratic countries, respondents should be selected with the same probability regardless of age, gender, ethnicity, etc.

Area Sampling: Sampling Without a Sampling List

Generally, in nationwide sampling, for example, two-stage sampling, a certain number of districts (areas) are extracted in proportion to the population of each district in the first stage. The second step is to find a fixed number of respondents (e.g., 10 or 20) at each district. If a sampling list, such as a list of residents or a list of voters, is not available, you need to consider how to sample respondents in each district you choose. Then use the residential map, if any, for each of the selected districts. If not, for example, walk around each district to create a map, as we did in 2002 Beijing Survey. In each district, households are selected by random-route sampling or the like, and individuals are selected from the selected households by the birthday method or Kish method. It is common not to extract more than one respondent from each household. This way of sampling using maps is called area sampling. Random-route sampling is a particular procedure of area sampling.

In the method of equal probability sampling of households such as area sampling, it is necessary to determine in advance a *weighting adjustment* that equalizes the sampling probabilities at the individual level. In this sense, sampling design is closely related to such weight adjustment. On the other hand, after the data is acquired, *weight adjustment* may be applied to the data to fit the population distribution of census data. Such adjustments are effective when the survey response rate (cooperation rate) is sufficiently high. However, in that case, the difference before and after the adjustment is relatively small. On the other hand, if the response rate is low, such weight adjustment may promote bias. For example, data that was originally biased due to low cooperation and low contact rates can be further biased in unpredictable directions (Note 2). Therefore, as long as the original sampling design is exact equal probability sampling at the individual levels, we often process the data without such weight adjustments. It is important to note that "bias" in the original data may contain some important information about the state of the country. Survey rejection rates or accessibility to respondents may be related to a country's economy or politics.

3.1.3.2 Face-To-Face Survey and Survey Culture

In many countries, it is not always easy to obtain statistical survey data that accurately represent the total population. In some countries, political instability has made nationwide surveys impossible, and it is necessary to infer national statistics from data of some cities and some rural areas.

In the United States, quota sampling and random walk sampling (essentially the same as random-route sampling) are commonly used in face-to-face surveys. However, since the late 1980s, due to cost and safety considerations, most survey firms have switched to telephone interview surveys, or more recently online surveys. Some firms may, if necessary, organize a network of local survey firms to conduct a nationwide face-to-face survey. We should be careful of the coverage, because many US national surveys exclude Alaska and Hawaii for technical reasons or costs.

Quota sampling in the United States creates a demographic quota table for gender, age group, ethnicity, etc., and the size of each demographic sample is proportional to the corresponding population. Even when there are only three demographic variables to consider (gender, age, and ethnicity), it is not easy to make the quota table exactly to match census data by triple-cross tabulation of variables, rather than each variable individually. Therefore, it is very difficult to consider more than three demographic variables. In addition, if you need to consider the statistical coherence of demographic distributions at both national and regional levels (sampling districts), it becomes insurmountably difficult. For example, if the percentage of Japanese Americans in the United States is 2%, then, e.g., 20 Japanese Americans need to be selected as part of a total sample of 1000 respondents. If 20 Japanese Americans in New York are selected as the sample, the sample is justified as national statistics. However, because the West Coast is home to a significant number of Japanese Americans, the sample does not match regional statistics. On the other hand, when sampled by regional population density, Japanese Americans may be unsampled or much less than statistically expected. This is because the pre-determined number of respondents in each district (e.g., 10 respondents) cannot include each minority that is less than 10% of the regional population density in most sampling districts. (This is the effect of rounding error.) To avoid such under-sampling, you might think that you should use the "Asian Americans" category instead of "Japanese Americans." Perhaps that is an idea. However, another issue can arise regarding biasing subcategories within a category.

In our 2006 US survey, research firms used a simple method to match gender, age group, and ethnicity assignments with US-level census data; thus, regional population distribution was not considered. This is the usual case.

In some countries, such as **Switzerland** and **Austria**, phone numbers had to be registered in the phone book, allowing statistical random sampling by phone survey. However, this is not possible in recent years due to privacy protections (enforced by the General Data Protection Regulations [GDPR] in 2018) and the rapid proliferation of mobile phones. (Note that the distribution of valid phone numbers is usually not uniform across the range of possible numbers.)

In **mainland China**, the government, public security, and resident committees maintain their lists of residents, but they are usually not available for polls and social surveys. Our 2002 Beijing and Shanghai surveys used area sampling, limited to urban areas. We made a map of the sampled districts (the sampled residential areas), and sampled households with a three-door spacing. Unfortunately, the survey had to be conducted in a very short period of time and the cooperation rate for the survey was low. The target number of respondents was set to 1,000 and the number of visits continued until the target number was reached. For example, in Beijing, the interviewers visited 3634 households and received 1062 respondents. This rate is similar to the rate of random-route sampling in Europe, which was previously mentioned for the 1992 survey in Italy and the 1993 survey in the Netherlands.

In **Taiwan**, resident registration was institutionalized in 1906 under Japanese occupation. However, according to the 2002 Taiwanese survey firm we contracted with, the phone book was closer (but not perfect) to the actual situation than the resident registration. The situation may have changed further due to the recent spread of mobile phones and recognition of privacy protection.

The **Singapore** government effectively manages people's homes (mainly high-rise apartments) under one-party control, avoiding ethnic conflicts and discrimination. A list of households from the Statistics Bureau of the Ministry of International Trade and Industry can be used for the survey. Households are categorized by living style, such as high-rise condominiums and detached houses. A similar list is available in **Hong Kong**.

GALLUP **Korea** usually uses random walk sampling, often combined quota sampling with age and gender. Although statistically undesirable, survey firms in many countries use preliminary samples. This is especially true in urban areas where cooperation is too low for a preselected set of sampling districts. Therefore, you should pay attention to the true response rate or cooperation rate when analyzing data.

Thus, in East Asia, a statistically less problematic face-to-face random sampling survey can be conducted at the national level in **Japan, South Korea, Taiwan, Hong Kong,** and **Singapore**. However, keep in mind that questions about politics and religion are restricted in many of these countries except Japan.

In **India**, different agencies conducted our national surveys in 2008 and 2013. The former used "the Right-Hand method" (random-route sampling and the birthday method), and the latter used the resident registration list at the final stage of respondent sampling. India is a typical multilingual (over 20 official languages and over 1000 local languages), multireligious, and multiethnic societies. Although it may be statistically desirable to match various attributes such as social class (cast) and language between interviewers and respondents, it is difficult to do so. It is never easy for foreigners to understand the depth of Indian society (Perhaps even among Indians, it seems difficult to understand each other's situation across social classes). At best, what we could do was to do our best to prevent the arbitrariness of the interviewers. I think the research company did its best in its own way. But to be honest, I'm not necessarily sure about the quality of India's survey data. This may be due in part to my lack of knowledge and experience in research in India (see Note 3).

Here some sampling policies from world-renowned surveys are briefly mentioned. Dozens of countries participate in the World Values Survey (WVS) started by Inglehart et al. (University of Michigan). The policy is general regarding interview methods, sample size (1200+), and age (18+). Interview surveys can be conducted by reading the questionnaire or using computer-aided personal interview (CAPI). Area sampling is used in the General Social Survey (GSS) by NORC in Chicago. However, due to changes in the survey environment, the details of the survey method have changed over decades.

Eurobarometer has played a politically important role in the postwar unification process of European countries. However, there are many variations in sampling, and many countries use the random-route sampling. On the other hand, European Social Survey (ESS) is a statistically ideal example. Their policy is to ensure strict statistical sampling and cross-national comparability of statistical surveys on different languages and sampling methods. Therefore, quota sampling and random-routes are prohibited. If sampling list is not available, area sampling with detailed specifications and careful procedures is strongly recommended (see guidelines [10]). See also [11, 48] for overseas sample survey methods.

You can also download a series of cross-national comparative survey reports by the Institute of Statistical Mathematics (https://www.ism.ac.jp/editsec/kenripo/index.html). For further relevant information, see the website: https://www.ism.ac.jp/~yoshino/.

Most importantly for technical aspects, even within the same country, survey firms and personnel have varying skills, which can have a significant impact on the quality of survey data, so you need to be aware of their capabilities. Taking all this into account, even if the distribution of public opinion varies from country to country, it is not immediately clear whether this is a real difference in opinion or due to sampling. This problem can be, at least partially, solved by multidimensional analysis (Sect. 2.3).

Survey Culture

Currently, the survey cooperation rate and contact rate with respondents are becoming lower and lower in most of countries. The cooperation rate is related to people's attitudes: (1) interest in the survey theme (benefit), (2) sense of social contribution by participating in the survey (significance), (3) burden of the survey (including privacy protection) [12]. The contact rate is related to people's lives such as diversification of lifestyles and working styles. If the cooperation rate is low even with the proper procedures and efforts of researchers, the rate itself may be closely related to some aspects of each society. Thus, cooperation rates and contact rates with respondents may represent some important information of a country or social situation.

Countries make policy decisions based on survey data collected in their own way. Each method should be evaluated not only in terms of statistical theory, but also in terms of history and political background. The survey culture of each country should be mutually respected.

3.2 Paradigm for Cross-National Comparison

3.2.1 Cultural Link Analysis (CLA) and Cultural Manifold Analysis (CULMAN)

In cross-national comparative survey, the main issue is the pursuit of cross-national comparability under differences in language, sampling method, cultural background, political system, religion, etc. Comparing completely different countries from the beginning would not be the best way to make meaningful comparisons in our type of questionnaire survey. By comparing pairs of countries (or groups) that have similarities, such as language or ethnicity, and identifying similarities and differences, a more meaningful statistical comparison can be revealed. By gradually connecting these comparison links (pairs of countries), the chain of links is expanded, eventually allowing global comparisons.

In our cross-national survey since the 1970s, Chikio Hayashi and Tatsuzo Suzuki developed this idea as "Cultural Link Analysis (CLA)" . Beginning with the comparison of Japanese living in Japan with Hawaiians of Japanese ancestry, the chain of comparison is gradually expanded and the question items are appropriately modified accordingly.

Similarly, longitudinal studies such as JNCS can consider a chain of time series comparisons. The same wording in a particular question does not necessarily guarantee long-term comparability. This is because decades of changes in the social environment can change the meaning of some questions. So we need to consider linkage of comparison over years. In addition, considering a comparative chain of certain survey theme or items, we could clarify multidimensional aspects of countries and societies.

Thus, we have the following three types of linkage of comparison.

(1) Longitudinal Linkage (temporal linkage)
(2) Cross-national Linkage (spatial linkage)
(3) Thematic Linkage (item-structure linkage)

More explanation is as follows.

Longitudinal Linkage (Temporal Linkage)

Longitudinal surveys such as JNCS can be considered as a series of links if each link is considered to correspond to a survey for each survey year (or era) (see Fig. 3.3). Each link is connected to adjacent links, and the entire links form a long-time chain. Each link also corresponds to each set of survey questions (items).

In general, it is important to continue using the same question items in a longitudinal survey. However, if the survey continues for several decades, the meaning of certain question may change. As a result, you may need to modify the wording of the item or replace it with another.

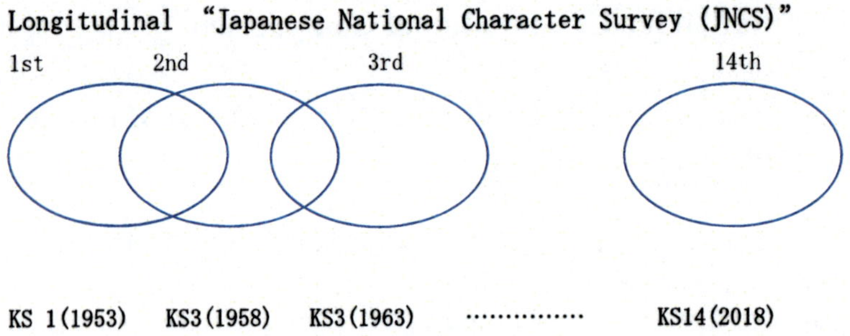

Fig. 3.3 CLA (Cultural Link Analysis): Longitudinal Linkage (Temporal Linkage). Each temporal local chart covers a specific time, and all charts together make a longitudinal linkage of time series comparison. Each local chart may correspond to a certain set of items under study

This is similar to the case of price index. A representative sample of daily necessities is the basis for calculating the price index, but it cannot be the same as it was 50 years ago. Some elements in the sample are replaced with others, e.g., every 5 years. For a certain period, both the old index and the new index are used in order to ensure continuity of longitudinal comparison. If necessary, certain weighting adjustments are applied to ensure continuity. By gradually modifying the sample set of daily necessities, it will be possible to connect short- or medium-term indices to make a long-term index.

In decades of longitudinal survey, adjacent pairs of temporal local links (temporal local charts) need to share the same item, but some items may be replaced by new items over time.

Cross-National Linkage (Spatial Linkage)

When trying to compare people's consciousness of countries or social groups that are far apart in terms of geography, culture, or religion, many questions arise as to whether the "same" question has the same meaning for different countries or groups. In consideration of this point, beginning with a chain of comparisons between two countries or groups that are assumed to have some commonality in important attributes such as culture, history, race or ethnicity, the chain (a spatial local chart) could be expanded to eventually form a global cross-national atlas (spatial atlas) (see Fig. 3.4).

Here again, each survey questionnaire (a set of question items) of each country/group corresponds to the spatial local chart. (This correspondence concerns the translation of the questionnaire).

Thematic Linkage (Item-Structure Linkage)

It is possible to compare the social structures of countries from various viewpoints represented by question items. On the one hand, certain items may only make sense

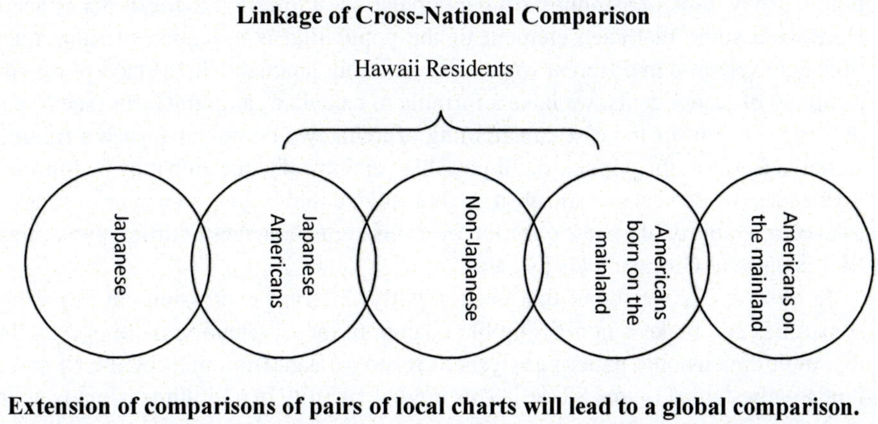

Fig. 3.4 CLA: cross-national linkage (spatial linkage)

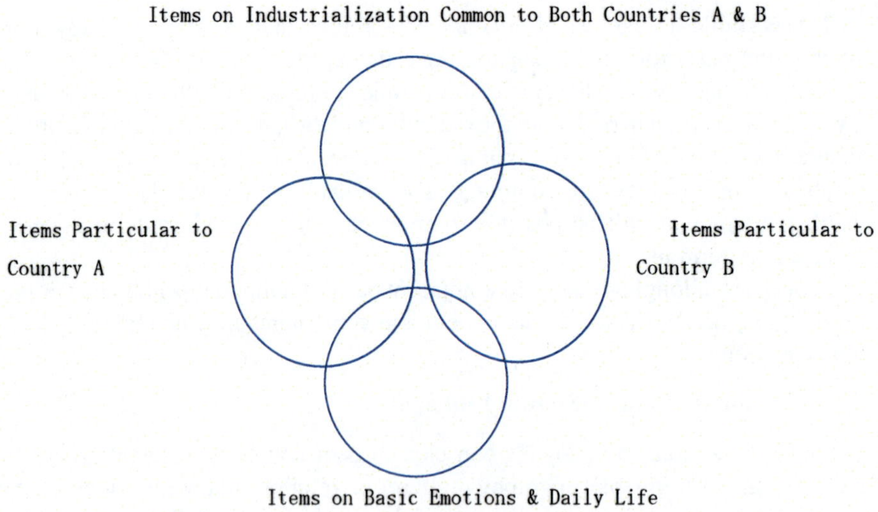

Fig. 3.5 CLA: thematic linkage (item-structure linkage). Relationships of similarity (commonalities) and dissimilarity are multidimensional. Each local chart of countries or regions under comparison may correspond to a set of questionnaire items or a set of topics under study

only for a certain country (or cultural zone). On the other hand, some other items may apply to more than one country (or culture zone). For example, items on basic human emotions may apply universally to all countries, while items on modern industrialization may only make sense in modern countries (see Fig. 3.5).

Each survey may focus on specific topics such as economy, industry, politics, and religion. If you find two survey that share a common item, you can create a link for comparison (a thematic local chart). Putting such links together form a multidimensional or multilayered structure.

On these three types of linkages, just like sampling of respondents from a population, we may think of sampling from a population of question items of our concern. Here we assume that each element of the population is a version of same question, but expressed in different wording or different language. In the case of random sampling of respondents, we have a formula to calculate sampling error (see Note 1 of Chap. 1). But for the case of sampling of items, we do not have such a rigorous formula. Besides, the population of possible versions of a question may be limitless. So the idea would be naive and abstract, but still it could be helpful when we consider as to comparability of a same question over different languages or different wordings (this is close to Guttmann [13]'s idea).

In Sect. 3.1.2, we found that even slightly different expressions in a question sometimes can make a non-negligible change in the distribution of responses, but that multidimensional pattern analysis of response data from multiple items makes it relatively stable [14, pp. 88–89]. This kind of stability in multidimensional pattern analysis can be thought of as corresponding to the formula for random sampling of respondents (see also Sect. 3.2.3).

Complementarity Principle in Survey Data Analysis

In our cross-national research, a kind of "complementarity principle" has been recognized in the following multiple senses (cf. Yoshino and Khor [15]).

(A) **Complementarity Between Theory and Practice**: Practical methods of data collection (survey methods) and theories (statistical theory of sampling error) mutually support and justify each other.

(B) **Complementarity Between Comparability of a Question Items and the Range of Countries (social groups) under Comparison**: The degree of comparability of a question item is affected by the range of countries compared (The point is, "Which would you like to see a specific tree or the whole forest?").

(C) **Research Theme Complementarity**: Focusing on a single theme for detailed analysis, or comprehensively researching on various themes.

(D) **Complementarity in Statistical Scaling**: One-dimensional scaling versus multidimensional scaling, or scope versus sensitivity (scale that comprehensively compares a wide range of countries, or scale that accurately captures some limited countries) (The strictness of the scale changes depending on the level of the hierarchy in Fig. 3.6. The higher the level of the hierarchy (that is, the wider the scope), the less stringent the comparability). And so forth. All these (A)–(D) would be intermingled.

Data analysis on complex real world would necessitate a certain theoretical basis. And theory must be verified and refined by practice. If we could always carry out a wide range and deep detailed analysis, there would be no serious problem in cross-national research. There is, however, usually a trade-off of proper balancing on the range of research and depth of analysis, as described in (A)–(D).

Cultural Manifold Analysis (CULMAN)

We have further expanded the scope of the cross-national comparative survey and introduced a hierarchical structure in the comparative links of these space, time, and item-structure as a paradigm of "Cultural Manifold Analysis (CULMAN)" [14, 16, 31, 32, 49–52] (see Fig. 3.6a–c).

Each link in a chain of each type of CLA (1)–(3) can be considered a local chart corresponding to a set of question items: a temporal local chart, a spatial local chart, and a local chart of item-structure. For each of the three types of chains, some charts may overlap or may indicate an inclusion relationship. Therefore, for each type of chain, the entire set of charts may create a hierarchy as an atlas. That is, in longitudinal chain, there are overlapping and inclusive relationships depending on the width of the period in which each survey is supposed to cover. This yields a hierarchy of temporal local charts. In the spatial chain, corresponding to geographical range, there are overlapping or inclusion relationships. This generates a hierarchy of spatial local charts.

Similarly, for a chain of item-structures, there may be a hierarchy of sets of items (or topics) depending on the overlapping or inclusive relation. For example, suppose

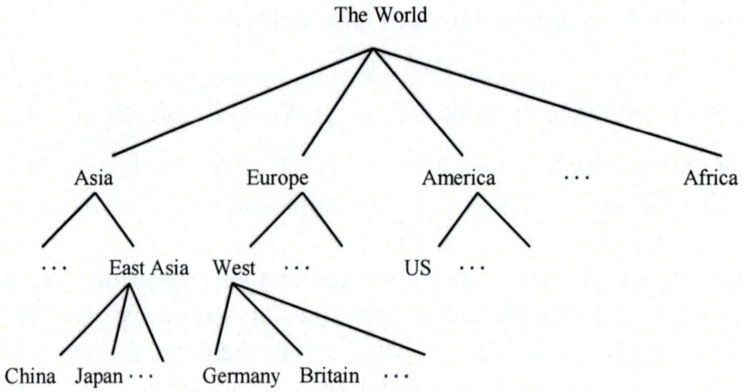

(a) Hierarchy of Spatial Local Charts

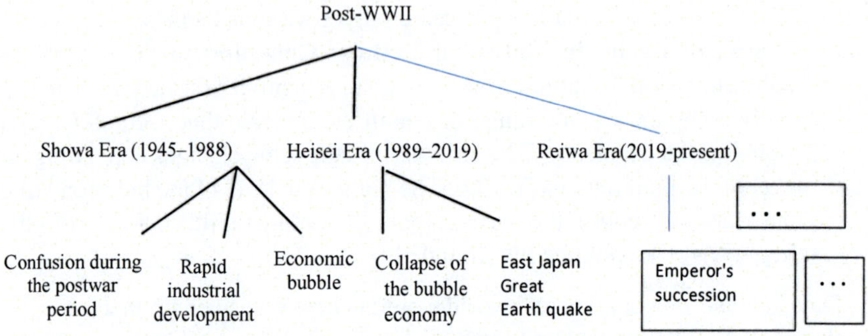

(b) Hierarchy of Temporal Local Charts of Japan

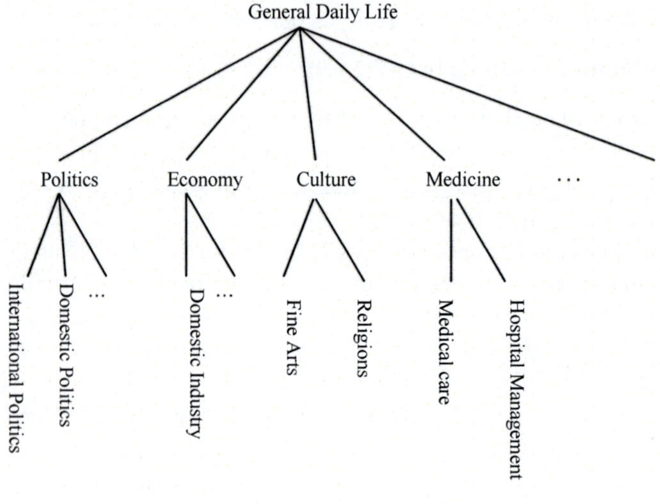

(c) Hierarchy of Local Charts of Item-Structure

Fig. 3.6 Cultural manifold: hierarchy of local charts (links). Some local charts may overlap each other within the hierarchy, and the structure may show dynamic changes over many decades that an item set A examines "industrial modernization" and an item set B examines "science and technology." The sets A and B might have some common items as well as some other items particular to A or B. From a certain perspective, set A may contain all items from B, or vice versa. Therefore, these sets can define overlapping or inclusive relations in a hierarchy, as in the case of set theory in mathematics.

This idea is expected also to serve as a reference paradigm for international policymaking (see Fig. 3.7) [17, 18], which will be discussed more in Chap. 5 and Chap. 6.

Connections of Local Charts for Expansion of Comparability

In CLA, connection of charts is important for smooth chain expansion (Fig. 3.3, 3.4 and 3.5). For example, in the longitudinal survey over decades, the survey items

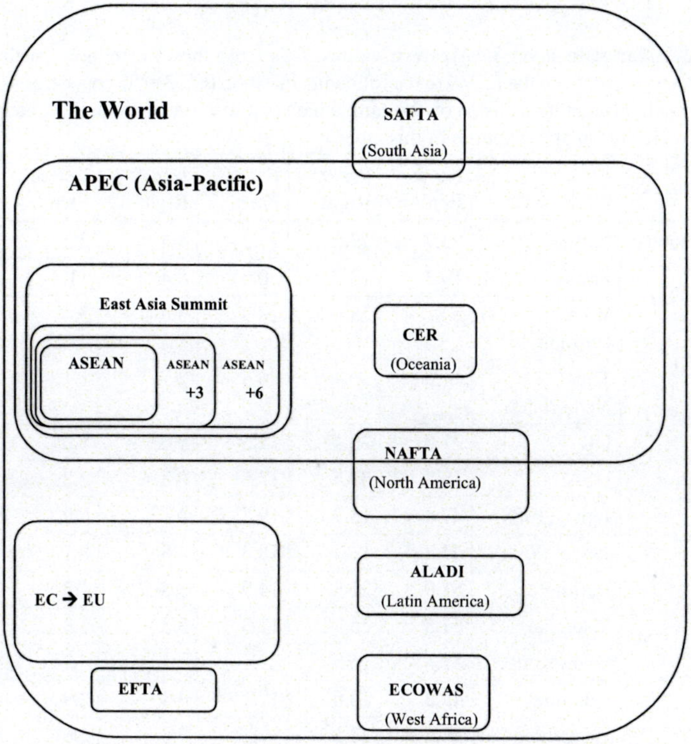

Fig. 3.7 Global manifold of local communities (an illustration). Each local chart (local community) may show dynamic changes in international relations. A new local chart may appear whereas another local chart may disappear. Two local charts may merge to make a larger local chart. Several pairs of local charts may overlap, and the entire set constitutes a hierarchy as a global manifold. Maintaining stable, peaceful and prosperous developments may require a set of "soft" regulations to connect pairs of local charts rather than a single restrictive global standard. The leaders of the modern world need to be intermediaries to resolve regional conflicts through global cooperation and harmony, not by military forces. (Recently, this manifold shows dynamic changes in local charts such as TPP, RCEP, etc.)

must be renewed, if necessary, without losing continuity as a measure. For chain expansion of the cross-national survey, two local charts (two sets of survey countries) must include, at least, a common region/country, as a key connector. For example, our Seven-Country Survey (1988–1993) and the East Asian Values Survey (EAVS) (2002–2005) both include Japan (1988 and 2001) (Table 3.1. If both Japanese data show the same response pattern for a particular item, it could be the basis for linking the two surveys and comparing all countries (i.e., at least tentatively you could assume the comparability of all the countries, disregarding time differences). See Table 3.2 and Fig. 3.8 as an example (This will be discussed more in Sect. 5.2 of Chap. 5).

However, if the Japanese response patterns of these two studies differ significantly, the time difference between the two surveys cannot be ignored. If this is the case, then do not immediately merge these two surveys. First, you should consider the results **of each surveys separately. Then, you should look at all the findings of two surveys**

Table 3.2 Asian values versus Western values. This table shows response distributions in the Seven Country survey and the EAVS to the following question (Q.33): "If you were asked to choose the two most important items listed on the card, which two would you choose? (Categories a-d for the options are shown at the bottom of this table.)

		Par & Ben	Par & Right	Par & Free	Ben & Right	Ben & Free	Right & Free
Seven Country Survey (1987–1993)	Italy	23.7	**29.2**	26.3	4.4	2.5	14.0
	France	19.3	13.7	20.4	8.4	11.4	**26.8**
	West Germany	8.8	28.0	19.3	3.6	3.4	**37.0**
	The Netherlands	11.6	**30.1**	26.3	2.7	1.1	28.2
	UK	**31.8**	20.4	12.5	11.3	8.0	16.1
	USA	19.5	**39.4**	12.5	6.3	2.8	19.6
	Japan (1988)	**47.4**	13.7	19.5	4.8	6.9	7.8
EAVS (2002–2005)	Japan (2002)	**47.2**	13.5	18.8	4.8	8.1	7.6
	Beijing	**51.0**	21.8	14.9	3.8	1.8	6.6
	Shanghai	**51.5**	25.7	12.3	3.3	1.3	5.8
	Hong Kong	**35.6**	16.3	29.1	4.1	5.7	9.2
	Kunming	**49.8**	20.6	17.1	4.4	2.9	5.1
	Hangzhou	**48.6**	22.5	16.1	4.5	2.6	5.8
	Taiwan	**56.7**	19.5	14.2	3.2	2.8	3.6
	South Korea	**45.2**	21.7	19.2	4.0	3.4	6.5
	Singapore	**50.3**	25.5	15.2	2.9	1.2	4.8

a Filial piety/love and respect for parents
b Repaying people who have helped you in the past
c Respect for the rights of the individual
d Respect for the freedom of the individual" (In the table, the categories are abbreviated as "Par," "Ben," "Right" and "Free," respectively)

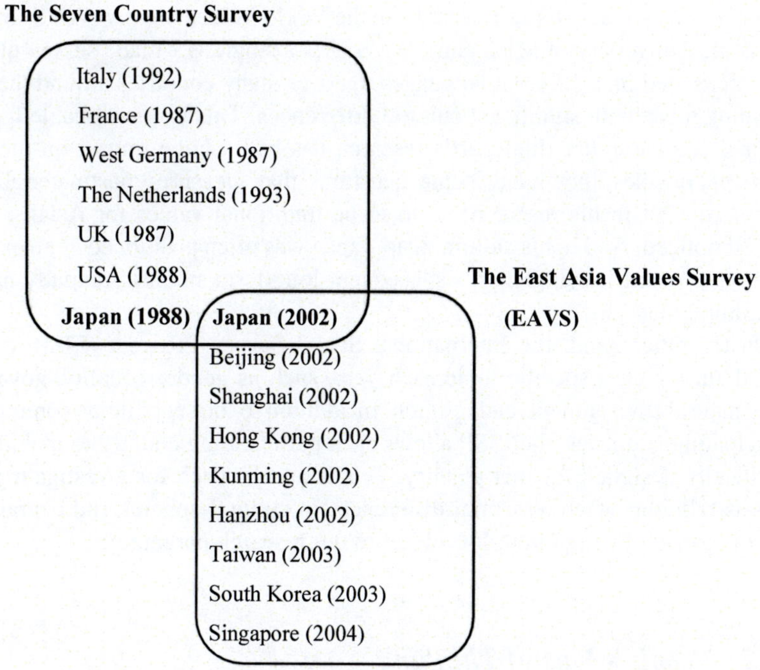

Fig. 3.8 Connection of two local charts: the Seven Country Survey (1987–1993) and the EAVS (2002–2005). Japan showed almost the same distribution of responses in both 1988 and 2002 surveys on a specific item of morality [18]. Combining these two local charts into a larger local chart, the comparability of time and country can be expanded under the assumption that the response distribution of this item has been stable for decades across the countries being compared

inclusively, taking into account the time difference (see Note 5 for the background idea of local chart connections).

The practical problem is how to handle the response distributions of two connected charts A and B to compare all members of the unified chart A ∪ B on the same scale. That is, we need to consider in a sense the relationship between the scales of Chart A, Chart B, and Chart A ∪ B. To solve this problem, it is necessary to have a proper design of question items and data analysis method that can obtain stable response patterns under various conditions in cross-national research. This applies to all three types of chains.

In a cross-national comparative survey, detailed scaling can be made for a comparison of countries in a certain limited area. But, if more countries are involved, then the scaling would inevitably become more coarse. In this sense, there is a kind of trade-off between accuracy and coverage on scale. Therefore, we need *a moderately sensitive and reasonably insensitive scale item design and data analysis method* that takes into account the range of target countries. Chapter 4 presents some examples on this issue.

Our survey paradigm is different from the World Values Survey (WVS) by Inglehart et al. (University of Michigan). WVS uses a single US-made survey questionnaire translated into different languages to cover many countries around the world, including those with significant cultural differences. This is questionable for cross-national comparability. Inglehart's research has had a huge impact on the world. However, on the other hand, some questions that measure "postmaterialism" (a central concept in his research) seem to be traditional values for Asians, as [19] himself noticed. And in his cultural map, Japan was often positioned as an exception from other countries (Inglehart's subsequent long-term studies gradually modified his cultural map [20]).

On the other hand, the International Social Survey Program (ISSP) conducts annual survey on a specific topic each year, such as gender equality, government roles, natural environment, and religion. In addition to the set of items common to all participating countries, the ISSP allows each participating country to include items specifically designed for that country. This is an approach for investigating topics such as religion, where essential differences in cultural spheres and overall social structures cannot be ignored. It is closer to our research paradigm.

3.2.2 People's Way of Thinking

Cross-Tabulation of Items

The 1971 survey of Hawaiian residents of Japanese descent ("Nikkei") was the first overseas survey conducted by the Institute of Statistical Mathematics. Not surprisingly, it was expected that some important differences would be found between Japanese Americans in Hawaii and Japanese in the homeland. However, the survey was seemingly unsuccessful when the distribution of answers to each question was compared with no remarkable difference. But Chikio Hayashi believed that despite sharing ancestral connections, he could find some important differences between people in these two societies. And he spent a lot of time analyzing the data in various ways and sought something important. Finally, he discovered a deeper level of data analysis.

His main idea is illustrated as follows.

For simplicity, suppose there is no difference between the Japanese and the Japanese Hawaii residents (Japanese Americans) with the Yes/No ratio (50%:50%) on both Question 1 and Question 2 (See Table 3.3). Furthermore, suppose that all Japanese who answered "Yes" in Question 1 answered "Yes" also in Question 2 and all Japanese who answered "No" in Question 1 answered also "No" in Question 2. As for the Japanese Hawaii residents, suppose that all those who answered "Yes" in Question 1 answered "No" in Question 2 and all those who answered "No" in Question 1 answered "Yes" in Question 2 (See Table 3.4). These response patterns are consistent with simple tabulation on each question (Table 3.3). This means that

Table 3.3 (Hypothetical example) Simple tabulation. Japanese and Japanese Hawaiians show the same answer distribution for each question Q1 and Q2 separately

	Japanese	Japanese Hawaii residents
	Yes : No	Yes : No
Q1	50% : 50%	50% : 50%
Q2	50% : 50%	50% : 50%

Table 3.4 (Hypothetical example) Example of cross tabulation of Q1 and Q2. These are consistent with the simple tabulations in Table 3.3, but show quite different response patterns for Japanese and Japanese Hawaiians

		Q2		
		Yes (%)	No (%)	Row total
Japanese				
Q1	Yes	50	0	50
	No	0	50	50
Column total		50	50	
Japanese Hawaii residents				
Q1	Yes	0	50	50
	No	50	0	50
Column total		50	50	

even if the distribution of answers from these two societies is exactly the same in each question, there can be a big difference in cross-tabulation of two questions.

In every country, there are subgroups that have various ways of thinking. If you compare two countries only by the percentage of responses to each of questions, the differences may not be so large compared to variances of response percentages of subgroups in each country. However, when comparing certain corresponding subgroups (e.g., gender, religion, etc.) of the two countries, response patterns ("the way of thinking") might happen to be quite different, and each country show each unique pattern (An example of a religion given by Fumi Hayashi et al. [7, p. 311]) is that the social values and consciousness of Catholic and Protestant believers corresponded in reverse between Europe and the United States).

Assume the following situation. The majority of country A (or culture zone A) answers questions Q1 and Q2 as "yes, yes" or "no, no." And the majority of country B (or culture zone B) answers these questions with "yes, no" or "no, yes" (These are close to the situation in Table 3.4. In addition, all people of country C (or culture zone C) answer these two questions without correlation between the questions. This means that for people in countries A and B, the two questions are linked to each other, but in different ways. But, for the people of country C, these questions are completely independent. If this is the case, regardless of how the distribution to each of these questions (percentages of 'yes' and 'no') is similar or dissimilar among countries A, B, and C, multidimensional structures (cross-tables) can distinguish between essential differences or similarities in countries (or cultural zones). Chikio Hayashi called such a difference "a difference in way of thinking." Even with different question sets, our cross-national data analysis often display clusters for a particular set of countries (e.g., the pair of UK and the United States, or the pair of France and Italy). This may reflect "the way of thinking" of each country and culture.

Crosstabs can be considered for any pair of question items. For example, if you have 50 questions in your survey, you may need to see a total of $50 \times 49 \div 2 = 1225$ crosstabs. In addition, you can consider cross-tabulating multiple items. Therefore, the total of crosstabs can be huge. It will be very difficult and time-consuming. Considering all, it is much easier to use multidimensional data analysis such as Hayashi's Quantification Method III (QMIII) ([6, 43, 46]), correspondence analysis [21], dual scaling [22], and optimal scaling [23] (All of these methods are mathematically equivalent but developed independently in their own disciplines).

Thus, we have come to use QMIII as a means of dealing with people's way of thinking (multidimensional structure of responses).

3.2.3 Hayashi's Quantification Method III (QMIII)

3.2.3.1 QMIII for Cross-National Comparison

As mentioned in Sect. 3.1.3, in cross-national comparative surveys, we cannot avoid situations where each country uses its own traditional method of sampling. In addition, some survey researchers or firms may modify response data by weighting, such as age, gender, and ethnicity, depending on census data. It raises the question of whether such weightings are always valid. Multidimensional analysis may give a specific solution to these problems.

Section 3.1.2 described the robustness of multidimensional analysis for slightly different wording of questions. Furthermore, Hayashi [4] and Yoshino and Hayashi [24] showed that one can disregard differences in sampling methods in a total configuration obtained by multidimensional data analysis (QMIII) when comparing data from many countries with respect to a group of items in contrast to an examination of only a single item. In addition, Yoshino showed that one could even detect falsified data by applying multidimensional scaling, called the "super-culture model" [61, 25, 26] (The model is closely related to the development of Cultural Consensus Theory (CCT) by Romney et al. [27] or Batchelder and Romney [28], and was influenced by my experience as their reseacrh assistant, where I contributed to the early stages of its development. For other related models , see [60, 61].)

Multidimensional analysis also works to solve the weight adjustment problem mentioned for the sampling probabilities of respondents. As an example, take our 1991 Japanese–Brazilian Survey. In Brazil, even among Japanese–Brazilians, there are many people with a mixed race of various kinds. Attempts to accurately adjust the data collected for racial combinations can significantly increase the relative adjustment weight of the data for a particular respondent [29]. However, on the contrary, such adjusted data can be significantly biased in terms of non-racial factors. As for the problem of weight adjustment of data, see Fig. 3.2 again. The figure is the output of QMIII for data of a certain set of question items from several countries/regions. Here we need to focus on two sets of data: weighted sample data from the Japanese–Brazilian (JBW) survey with respect to ethnicities and non-weighted sample data (JBS). As for Brazil's data, as mentioned earlier, there was a great deal of bias among Japanese-Brazilians due to the complex ethnicities that made it difficult to find the right weight adjustment. The weighting coefficients could be large enough to make the data less reliable (i.e., the variance could become too large). However, in the QMIII output, the difference between the two datasets (JBS and JBW) is so small that the effect of weighting can be ignored in response pattern analysis when comparing data for multiple items in many countries.

Thus, multidimensional analysis may offset differences in item wordings, sampling methods, and weight adjustments, to provide a stable macro pattern (Yoshino et al. [51, Sects. 2 and 3]). Multidimensional data analysis provides a consistent overall analysis, losing some of the details of individual item data. This is one of the "complementarity" mentioned earlier in this section.

3.2.3.2 Unidimensional Scale and QMIII

For scaling on survey data, we need to pay attention to the characteristics of scale type (nominal, ordinal, interval, or ratio scale in the S.S. Stevens' classification [30, 40]), of data. The main concern here is interval scaling of ordinal data using multidimensional analysis such as Hayashi's QMIII.

For example, life satisfaction questions are often asked on a set of choices, such as "1. Satisfied, 2. Somewhat satisfied, 3. Somewhat dissatisfied, 4. Dissatisfied" or a 7-point scale. However, the superficial "numerical value" of choices may not make sense on an interval scale. For example, the psychological interval between "choices 1 and 2" is not necessarily equal to the psychological interval between "choices 2 and 3." This is also the case with 7-point scale. Furthermore, in some cases, those choices may not make even an ordinal scaling, as we will see later. Strictly speaking, these are just categorical data (nominal scale) unless there is more justification to the scale level of the data.

It is important to recognize the distinction between the intended scale level when collecting data and the scale level when analyzing data. In the data analysis of categorical data (nominal scale), one may have options such as (1) application of analysis suitable to categorical data, or (2) interval scale construction on multidimensional analysis. An example of the latter method is outlined below.

When applying a multidimensional analysis such as QMIII to data asking for "life satisfaction," "health satisfaction," or "environmental satisfaction" under the previously shown four choices, we often get a pattern like Figs. 3.9a or b (The same can be applied to the 7-point scale). Regarding the three types of satisfaction, the

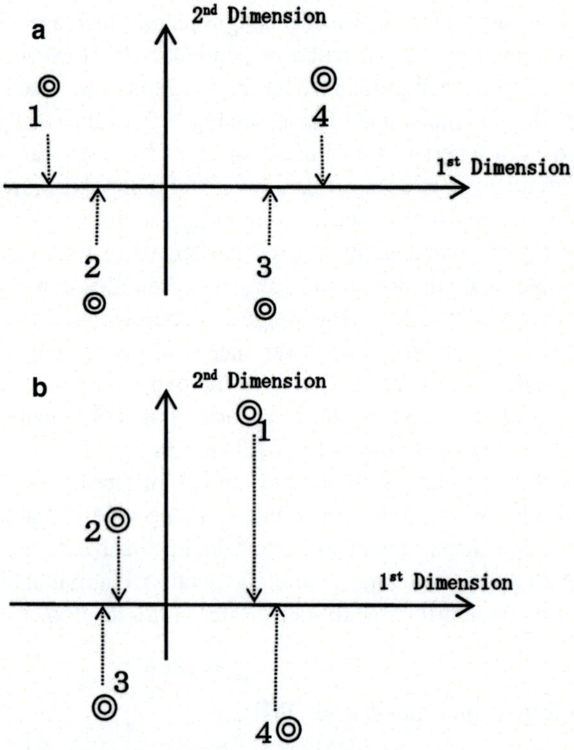

Fig. 3.9 **a An illustration that choices 1–4 follow a U-shaped distribution.** The locations of these choices projected onto the 1st dimension (i.e., the first coordinates) keeps the ordering of Choices 1–4 ("Response Choice 1" of each item on "satisfaction with daily life, heath, environment, life in general" often appears near each other. The same is true for the other choices). **b An Illustration that Choices 1–4 Follow a ⊂-Shaped Distribution.** Firstly, two groups of Choices 1 and 4 (the polar answers) and Choices 2 and 3 (mild answers) are separated along the 1st dimension (right and left). Then, along the second dimension, choices 1 and 4 are separated into positive and negative (upper and lower), as are choices 2 and 3. Thus the first coordinates of the polar responses choices 1 & 4 are close to each other, so are the mild response choice 2 & 3

patterns of response choices often overlap each other. For illustration purposes, these figures show the choices for only one question. The similarity of a pair of choices as response data corresponds to the distances between the pair in these figures.

If the result is U-shaped as shown in Fig. 3.9a, it is possible to proceed to the judgment of ordinal scale.

(1) Confirmation of Ordinal Scale

If the first dimensional eigenvalue (corresponding to the "explanation rate" of the first dimension of data variance) is sufficiently large compared to the eigenvalue of the second and subsequent dimensions, and choices 1–4 are arranged in order on the first dimension, the set of choices can be considered an ordinal scale.

In general, the order of the first dimensional coordinates of the choices "1," "2," "3," "4" sometimes can differ from the superficial order (although this may be unlikely to happen on the question of "satisfaction"). For example, the order might be 1, 2, 4, 3. If the first dimensional eigenvalue is not large enough compared to the second dimensional eigenvalue, then you should consider whether to process only the first dimensional components or include the higher-dimensional components. If you decide to handle only the first dimensional components that make up the ordinal scale, you may further check whether it can be an interval scale.

(2) Interval Scale

Check if the first dimensional coordinates of each choice (category) can be considered as being evenly spaced. For example, suppose the four choices of the first dimensional coordinates are 0.98, 2.11, 2.98, and 3.89, which are approximately evenly spaced. The choices can be viewed as making a crude interval scale. For brevity, the choices can be considered as approximate scores 1–4. For another example, in the first dimensional coordinates, if the spacing between choices 1 and 2 is 1.1, the spacing between choices 2 and 3 is 1.9, and the spacing between choices 3 and 4 is 0.95, then approximate scores of the choices can be assigned as 1, 2, 4, 5 on the interval scale. Of course, in these examples, the first dimensional coordinates of QMIII (not necessarily integers, but generally real numbers) can be used as the exact scores of the interval scale. It would be a practical decision whether to use the exact coordinate or the approximate integer for simplicity. Calculations such as the arithmetic mean are statistically meaningful on the interval scale, but not on the ordinal scale.

Interval scales constructed in this way can look like Likert scales that use the arithmetic mean of selected superficial numbers ("1–4"). But the Likert scale does not necessarily have a priori theoretical justification as an interval scale. In a sense, our method modifies and justifies the Likert scale as an interval scale.

(3) Multidimensional Scaling

However, we often get a ⊂-shaped pattern shown in Fig. 3.9b. In this case, first, the two groups of extreme choices ("1" and "4") and the two groups of middle choices ("2" and "3") are separated in the first dimension, and the second dimension is divided into positive (satisfied) side and negative (dissatisfied) side within each of the two groups. For example, a scale of rightist and leftist in political attitude is often displayed at the two extremes of the 10-point scale (i.e., "1" and "10"). However, an analysis of the extreme right and extreme left personality structures often reveals that they are very similar. It is not uncommon for radical left-wing activists suddenly to become right-wing activists and vice versa. This may be a phenomenon occurring not only in Japan but also in other countries.

In such cases, it would be misleading to simply analyze the data on superficial numbers of choices or on a 10-point scale. First, it is necessary to analyze the rich information contained in multidimensional data as is. Each choice is arranged in a multidimensional plane or space (as a Cartesian coordinate). And the data of mutual distances of choices in the multidimensional space may give further analysis such as cluster classification (Here, the positioning of response choices is described, but the positioning of each respondent or country could be described in the same manner).

Thus, multidimensional analysis can go beyond superficial scaling to find essential structures in the response data.

3.3 Procedure of Cross-National Survey: Designing a Questionnaire and Sampling, Pretest, Nationwide Survey, Data Cleaning, and Data Analysis

The procedure for our cross-national surveys generally consists of the following eight steps.

Step 1: Designing a questionnaire in a native language (e.g., in Japanese).
Step 2: Designing a questionnaire in a foreign language (translation and back-translation).
Step 3: Choosing a survey sampling method for each country.
Step 4: Developing a semi-final version of the questionnaire.
Step 5: Conducting a pretest survey based on a small sample to finalize the questionnaire.
Step 6: Conducting a nationwide survey (data collection).
Step 7: Data cleaning (including recoding response categories).
Step 8: Data analysis (cross-tabulation and pattern analysis).

These eight steps are closely interrelated and controlled through feedback to overcome problems in comparing cross-national survey data. The feedback process forms a spiral of longitudinal and cross-national research. For each step, some points to note are provided below. Here we assume that the original questionnaire was in Japanese, but the same applies in other cases.

Collection of Domestic and Overseas Research Materials

In Step 1, first and foremost, we need to collect information about past studies related to the item you want to investigate. List up all items or topics to be surveyed. If we find a related question in an existing survey, decide whether to use it for the survey as it is or modify it. Established surveys such as JNCS, General Social Survey (GSS) of the United States, French CREDOC, German ALLBUS, European ESS, ISSP (International Social Survey Program), etc., are available. If we can't find the appropriate question in existing surveys, we need to make a question ourselves, taking into account the details of the wording, answer categories (choices), and order of the newly created question to make it look natural and simple in the questionnaire.

However, we need to be careful about **objective statistical data collected from other sources**. For example, GDP, unemployment rate, college enrollment rate, religious affiliation rate, etc. should be collected from existing sources as objective data.

Even if these items are also included in our questionnaire, the objective information is necessary to check bias in our survey data.

If your question has too many or too complex choices, create a handcard of choices that will be displayed to respondents. Also, if the question itself is complex, write it down on your handcard and present it to the respondents when asking the question.

Creating a History List of Survey Questions

It is recommended to create a history list of survey questions, just as a doctor creates a medical record (chart). The list shows in which survey each question was used in the past, and what results were obtained (whether it was used successfully, failed, or the expression was changed in your survey, and if so, why, etc.). The analysis that can be done only from new data is very limited, so a comprehensive data analysis with information from other surveys needs to be performed. The literacy on survey data can be more or less determined by the richness of the content of the history list of survey questions. The list is an important asset for survey researchers (for an example, see pp. 58–102 of Yoshino et al. [31]: https://www.ism.ac.jp/editsec/kenripo/pdf/kenripo117.pdf).

Attention must be paid to the ethics and responsibilities of the interviewer. *Some delicate questions should not be asked or can only be asked by following very strict informed consent procedures.* For example, with the exception of Japan, many countries in East Asia and Southeast Asia cannot ask deep questions about politics and religion. Even if a survey is conducted with the consent of the respondents, the consent process can result in biased responses. Or simple questions about serious illness and death may leave trauma in the minds of respondents facing those problems. In APVS survey, we paid particular attention to a question concerning cancer notification (Q23), allowing respondents to easily skip it if they wished.

Some items cannot be investigated due to accidents, privacy protection, or for moral reasons. Ordinary public opinion polls do not allow medical examination of respondents (When conducting a medical examination, it corresponds to a public health survey based on the strict regulations of the Privacy Protection Act). Certain longitudinal surveys of nuclear power plants could not continue shortly after the 2011 Fukushima nuclear disaster because it was necessary to worry about people's trauma.

Designing Questionnaire in a Foreign Language

Once a set of questions are determined, it is necessary to translate them into languages of countries to survey. The semantic equivalence of the set of questions in different languages need to be verified by performing the translation and back-translation in step 2. However, no matter how you translate, even a slight difference in the wording of a question in the same language can make a non-negligible difference in the response. Therefore, verification of equivalence is not always possible. Section 3.1.2 showed that if we compare the response patterns of several countries for many of the items by performing multidimensional data analysis, we may be able to ignore the influence of minor differences in wording entailed in translation processes [24]. Therefore, it is advisable to analyze the data for a set of questions rather than as a single question. When designing your survey, we recommend that you create a set of questions related to each topic of interest.

Sampling Method

In each of the target countries, we signed a business contract with a reliable survey firm, sometimes with the help of academic researchers. If you need to work with academic teams in the target country, you need to verify their knowledge and ability with respect to actual statistical sampling surveys.

As mentioned in Sect. 3.1.3, many countries may apply their own established sampling methods also in cross-national comparative surveys. The robustness of outputs in multidimensional analyses is exemplified in Sect. 3.2.3 for sampling differences.

Pretest Survey and Nationwide Survey

Steps 3 and 4 are preparations for Step 5 to conduct a pretest survey with a small number of respondents, such as in urban and rural areas. Prior to Step 6 of the nationwide survey, if you find any problems on the questionnaire or sampling methods, you need to correct them at this step.

In the preparatory steps before the nationwide survey of Step 6, it is highly recommended to collect relevant information in each country as possible as you can. Don't be content with just waiting for the data to come back from the entrusted survey firm. For example, interview a few respondents on your own, or at least observe a few interviews with native interviewers in the target country. You can visit places such as schools and restaurants where you can find people who can speak openly about economic, political, and social situations of the country. You can visit international fast food restaurants and compare prices with those of your country. You should do your best to gather much information in such an informal way in order to help analyze the sampling survey data. All of these things are useful when assessing the quality of survey data and analyzing the data.

Data Cleaning

Data cleaning looks tedious work simply to contrast the questionnaire returns with a file of input data in a summary table or in digital device, etc. However, reviewing each response sheet of questionnaire is a professional task that helps you to understand how each interview was performed. For example, it may be possible to infer from interviewer's handwriting whether each question and answer took sufficient amount of time for the survey, or whether it was completed in a short time. Sometimes forged data may be found.

Multiple choice questions may show sometimes unexpected answers recorded as "Other answers [specific …]." In such a case, it may be better to create a new category in the summary table if more than 5% of all the respondents give essentially the same answer. Here, what is important is to see if there are many such unexpected responses in only certain countries, or if similar results were obtained in most of the countries surveyed. A large number of unexpected responses may be usually considered as a questionnaire design failure. However, it may represent important country differences in the nature of the meaning of the survey questions.

Data Analysis

First, create some tables: (1) a summary table of responses to each question, (2) crosstabs of responses to each question by each demographic variable (gender, age, occupation, income, education, etc.), and (3) crosstabs of pairs of specific questions of particular interest. You can also apply multidimensional scaling if desired.

We should pay attention when making cross-tabulation of certain item and demographic variables. As for variables such as educational level, income level, we usually need to re-categorize the choice categories of the variables to secure cross-national comparability. For example, with regard to income or education level, it is not easy to ensure comparability across countries simply by converting exchange rates of money or years of schooling. After understanding the details of each country's situation or system, response categories of those variables can be reclassified into, say, three levels (high, middle, and low) in each country. It may be an idea to classify the three levels of the variable according to the distribution of the response data, with each level corresponding to the bottom 1/3, middle 1/3, and top 1/3 of the entire sample. The ILO defines the International Standard Classification of Occupations (ISCO), but due to different circumstances in different countries, it is not possible to ensure complete comparability of occupation across countries. For example, "skilled workers" are true experts in Germany, but in other countries it may include workers who have been engaged in the same simple task for decades. Therefore, it is not always easy to compare these demographic variables across countries. But this difficulty itself can be an important clue for understanding each country.

On the other hand, in data analysis, it is necessary to pay attention to cross-national differences of general response tendency. For example, the Japanese tend to avoid clear "yes/no" answers and to prefer ambiguous or medium choices, the French tend to choose critical or negative choices, and the Indians tend to choose optimistic or positive choices [14, 36]. This warns us not to jump to a superficial comparison of the response data. This kind of general response tendency characterizes people not only at the national level but also at the individual level. Each nation is composed of different groups of personalities (cf. Sect. 2.4.9 of Chap. 2). Differences in the distribution of personalities (rationality, self-disclosure, etc.) represent important differences in national character between countries. Therefore, it is important in our analysis to distinguish between individual and national levels. More details will be shown in next section.

3.4 People's General Response Tendencies

3.4.1 Some Observations on Response Tendency

This section deals with the general response tendencies observed in our longitudinal and cross-national research [32, Sect. 2]. Here, the "general response tendency" means a common attitude particular to a specific country (or group) responding in a public opinion survey or a social survey, regardless of the item or topic in question. For example, French people tend to give negative or critical answers to any question [26]. This may mean they are pessimistic. Or, it may be closely linked to their critical attitudes as an aspect of mature democracy [33, p. 258]. Brazilians showed optimism comparable to wealthy Europeans even during the financial difficulties of the 1980s, as evidenced by a survey of happiness and life satisfaction [34].

Japanese people, on the other hand, avoid clear "yes/no" answers or polar categories (e.g., "very satisfied" or "very dissatisfied") and use moderate response categories (e.g., "somewhat satisfied" or "somewhat dissatisfied"). The response rates of "DK (don't know)" and "Undecided" are also higher than in other countries (Such tendency is designated as "middle-category response tendency [MCR tendency]" throughout this book) (Some researchers attribute this tendency to the Japanese emphasis on harmony since the reign of Prince Shotoku, who struggled to subdue the serious fighting of people in the sixth century). For example, even in the 1980s, when the Japanese economy was nearing the top of the world, the Japanese did not show high satisfaction in questionnaire surveys. This was amazing for the world (at least the west), but it can be explained by MCR tendency. Therefore, it is important to remember that survey response data is not simply a reflection of the actual situation but is complicated by the general response tendency in this sense.

It may be worth mentioning here a bilingual comparative study to show that even the same bilingual respondent may give you different answers to the same question if given in different languages. To investigate this issue, Hayashi and Suzuki [35] compared bilinguals of Japanese and English by split-half method on Japanese and American college students. They found that, when responding to Japanese questions in Japanese, both Japanese and American students tended to avoid extreme responses (e.g., "very...") and chose mild or vague response choices or "Don't Know." On the other hand, respondents tended to choose polar choices or clear Yes/No answers when responding to English questions in English. Moreover, the Japanese were as individualistic and optimistic as Americans when they were responding to a question in English (cf. linguistic relativity of Sapir & Whorf hypothesis).

In a more recent survey data from the United States, Australia, Singapore, and India, where English is a public language (or one of public languages), Yoshino [26, 36] suggests that in multiethnic countries, people tend to choose more extreme choices or give a clear "yes/no" answer [32, Sect. 2]. This can be due to communication problems using a "common language" (such as Arabic or English) rather than one's native language. This may explain the above-mentioned observations by Hayashi and Suzuki [35] about the Japanese tendency to respond in English.

Survey data from Australia, Singapore, and India may be useful for research from this perspective as these multiethnic countries use English as one of the official languages.

Incidentally, in a philosophical study, Umehara [37], a famous Japanologist, believes that the ambiguity of Japanese remarks and actions arises from the intermediate ambiguity of polytheism since the Jomon culture (some 16,500–2400 years ago).

Yoshino [32] analyzed in detail the differences in the response tendencies and gender differences of Japanese immigrants in Brazil, Hawaii, and the West Coast of the United States. It was concluded that the general tendency of responses in the questionnaire survey may be closely related to the degree of self-disclosure of each country or individual. However, when using a psychological scale such as a distress scale (such as the Kessler Psychological Distress Scale), it is necessary to carefully examine the differences in situations that may occur in individual response tendency. For example, if a relationship of trust is established between the patient and the doctor, the distress scale can be used effectively for the patient. On the other hand, if the same scale is used in a social survey where the interviewer meets an unknown respondent, such a scale can lead to paradoxical results. For example, Japanese men were shown to have the least pain on the pain scale in their 50s, but their generation had the highest suicide rate [14]. This seems to be closely related to the issue of self-disclosure (Japanese men tend to close themselves when they are suffering from severe pain). This will be explained again in Sect. 3.4.3.

3.4.2 Cross-National Difference of Choice of Middle Category

This sub-section compares cross-nationally the extent of MCR tendency in survey data. This includes not only comparisons between countries, but also comparisons between races of the same country and generations of Japanese immigrants. There may be many ways to measure MCR, but I tentatively use the sum of the polar response percentages in the next question (from APVS Q18) as the opposite rate of MCR. This question asks respondents to rate the importance of seven domains of daily life on a seven-point scale (1="not important at all" to 7="very important").

	Not important at all						Very important
a. Your immediate family members such as spouse and children, if you have any	1	2	3	4	5	6	7
b. Career and job	1	2	3	4	5	6	7
c. Free time and relaxation	1	2	3	4	5	6	7

(continued)

(continued)

	Not important at all					Very important
d. Friends and people you know	1	2	3	4	5	6	7
e. Parents, brothers, sisters, and other relatives	1	2	3	4	5	6	7
f. Religion	1	2	3	4	5	6	7
g. Politics	1	2	3	4	5	6	7

Q18 [SHOW CARD 12] Using a scale of 1–7, where '7' is "very important," and '1' is "not important at all," can you tell me how important each of the following is to you?

In this book, by definition, the sum of the polar responses "1" and "7" to all of the seven domains (a, b, c, …, g) is called the *rate of polar responses (abbreviated as RPR hereafter)*. I have analyzed the data with several possible statistical measures concerning RPR or rate of middle category response. It is difficult for any metric to separate the real intent of choosing very positive or very negative categories. Therefore, even with this RPR, it is not possible to fully grasp the tendency of polar response as a general response tendency, but for simplicity, I use RPR provisionally.

Tables 3.5, 3.6 and 3.7 shows RPR for all previous surveys that included the same seven sub-questions as Q18a-g. These show the results of the Seven-Country Survey (1987–1993), EAVS (2002–2005), PRVS (2004–2008), and Japanese Immigrant Survey in Hawaii, Brazil, and the West Coast of the United States. Some points in Tables 3.5, 3.6 and 3.7 are as follows.

(1) The patterns of RPR in the Japan A and Japan B surveys in 1988 were the same (Table 3.5), which may suggest the stability of RPR as a measure. The RPR pattern shown in the 2002 Japan survey as part of EAVS (Table 3.7) was

Table 3.5. The Rate of Polar Responses (RPR) of Q18a-g. The Seven Country survey

Figures show the sum of polar responses "1" and "7" to Q18a-g [seven subquestions].)

year	1988
	Japan-A
RPR	317
year	1988
	Japan-B
RPR	317

1988	1987	1993	1987	1987	1992
USA	UK	HOL	WEST GERMANY	FRANCE	ITALY
338	306	284	222	309	312

Table 3.6 The Rate of Polar Responses (RPR) of Q18a-g. Japanese Immigrant surveys

year	1999 Hawaii	
	Non-Japanese Americans	Japanese Americans
RPR	311	318

year	1988Hawaii	Japanese Americans		
	Non-Japanese American	Japanese Americans (total)	1st & 2nd Generations	3rd & 4th Generations
RPR	320	314	317	295

1991 JAPANESE BRAZILIANS			
total	1st Generation	2nd Generation	3rd Generation
457	514	463	413

1991 JAPANESE BRAZILIA (Degrees of Japaneseness)				
total	Low	Middle Low	Middle High	Middle High
457	465	413	465	413

1988 the West Coast of USA			
Japanese Americans (total)	2nd Generation	3rd Generation	4th& 5th Generations
319	380	327	300

Table 3.7 The Rate of Polar Responses (RPR) of Q18a-g. The East Asia Vaules survey (EAVs) and The Pacific-Rim Values survey (PRVs)

year	2002	2002	2003	2002	2002	2003	2003	2004	2002			
	Beijing	Shanghai	Kunming	Hangzhou	Hong Kong	Taiwan	South Korea	Singapore	Japan			
RPR	340	287	372	343	258	280	407	359	312			
year	2005	2005			2005	2006	2006	2007	2004	2007	2006	2008
	Beijing	Shanghai			Hong Kong	Taiwan	South Korea	Singapore	Japan A	Australia	USA	India
RPR	328	344			265	249	299	338	265	309	325	326

2006Taiwan		
total	Native Taiwanese	Chinese mainlanders
249	249	254

2004	Singapore			
total	Chinese	Malay	Indian	Others
359	327	431	435	393
2007	Singapore			
total	Chinese	Malay	Indian	Others
338	317	425	382	388

similar to those of Japan A and Japan B 1988 surveys. However, in APVS (2004), the RPR value was lower by approximately 50 points. It is not possible to immediately conclude how much RPR is inevitably affected by changes in opinion over time, and how the context of the study affects it.

(2) Throughout the past surveys, Japanese-Brazilians had the highest RPR (Table 3.6), and 1988 West Germans (before the integration of East and West Germany) had the lowest (The questionnaire for Japanese-Brazilians was in Brazilian Portuguese). South Koreans had the second highest value in 2003, but in 2006 it fell by more than 100 points. It is not clear whether these differences represent actual changes in public opinion or the variability of RPR.

(3) Japanese immigrants and their offspring in Hawaii, Brazil, and the West Coast of the United States had the lowest RPR in the younger generation. This is consistent with Hayashi's observations [7, p. 389] that these Japanese groups tend to lead to avoidance of polar categories and preference for the middle category. Kuroda and Suzuki [53] attribute response tendencies mainly to culture or language rather than to other types of social conditions. But our data suggest the effects of generational differences also.

On the other hand, Hayashi [7, p. 86] divided the respondents of the Japanese–Brazilian survey into four groups according to the degree of Japaneseness, but RPR was not proportional to the degree of Japaneseness.

Japanese immigrants on the West Coast and their children had severe experiences such as racism during World War II ([38, 39]). Minorities people need to develop attitudes that allow them to speak up to defend their rights and gain political power. In that case, we can infer that RPR tends to be higher in multiethnic countries and social groups. In fact, this is confirmed by our data on the United States, Singapore, and India, as well as for Japanese immigrants (Tables 3.5, 3.6 and 3.7). We also confirmed that the data from Hayashi et al. ([7, p. 389]) showed that Japanese immigrants in Hawaii had a much lower MCR tendency. As Kuroda and Suzuki explained ([53]), culture and language may be closely related. However, in multiethnic societies, people's ability to communicate is limited if they must use the same official language, such as

English. Therefore, people need to speak clearly to avoid misunderstandings. Therefore, RPR would naturally tends to be high.

(4) Even in English-speaking countries, there are differences depending on countries and ethnic groups. RPR is high in countries where English is one of the several official languages, such as Singapore and India.

(5) In Taiwan, there is little difference in the RPR between native Taiwanese (those born and raised in Taiwan) and "Mainland Chinese" (people who emigrated to Taiwan during the post-World War II political turmoil and their descendants). The slight difference in RPR may suggest less political tension on domestic policy between these two groups than before. But if so, it seems important to see if RPR could fluctuate in response to political tensions (see Note 6).

(6) There are three large ethnic groups in Singapore: Chinese (majority), Malay, and Indian (minorities). In RPR, Chinese were the lowest and Malaysians and Indians were much higher in both 2004 and 2007 (The Indian RPR was changed from 435 to 382, but the general pattern of RPR was consistent among the three ethnic groups). This observation also confirms that RPR in ethnic minorities tends to be larger in multiethnic societies.

(7) Regarding gender differences, RPR was generally higher in women than in men (For gender differences, see the demographic cross-tables in a series of our survey reports). This difference may be closely linked to the issue of self-disclosure, as shown in Sect. 3.4.3 on responses to self-reported medical symptoms [14, 32].

Many factors may be involved in response data. However, it is certain that responses in our surveys always involve some general response tendencies, particular to some countries or ethnicities. Moreover, such general response tendencies characterize people not only at the national level but also at the individual level [14]. This may be also related to the classification of "rational type" and "non-rational type," as described in Sect. 2.4.9 of Chap. 2. This means that each country is made up of groups with different personalities, and the differences in the distribution of personalities represent some differences in national character. Differences in national character may appear as differences in the distribution of personality types, not as differences in personality types (See Note 7 for an experimental study of response distributions in each individual's mind).

3.4.3 Degrees of Self-disclosure: National Differences and Gender Differences

Our survey of seven countries (1986–1993) targeted Japan, the United States, the United Kingdom, France, West Germany (before unification), Italy, and the Netherlands. Multidimensional analysis such as QMIII or Yoshino's "super-culture model" ([25]) found three clusters: (a) UK and the United States (common origin and language), (b) France and Italy (Latin nations and geographical neighbors), and

(c) Germany and the Netherlands (geographical neighbors). Disregarding of slight different choice of question items, multidimensional analysis of these countries often finds the same clusters. Perhaps, in our type of social survey data, clustering can be relatively robust, at least with regard to subtle differences in choice of question items regarding attitude, consciousness, and values.

Yoshino and Hayashi [24] gave an example of question item that initially seemed unrelated to national character but later turned out some relevance to national character. The question is as follows:

Q14. During the last four weeks, have you suffered from any of the following?
1. Headaches/migraines 2. Backaches 3. Nervousness 4. Depression 5. Insomnia.

Since this question was about a medical issue, it clearly seemed to have nothing to do with national character. See the data from the Seven Country Survey (1987–1993) on the left side of Fig. 3.10. The figure shows the average number of symptoms. The rate of people who claimed to be afflicted with these health problems was higher in women than men in all seven countries. Besides, the rate of Japanese men experiencing these problems was found to be the lowest [7, pp. 126–128] (This pattern is similar to the pattern of the proportion of people who claimed to have any of the five symptoms (Yoshino, Hayashi, and Yamaoka [14, p. 153]).

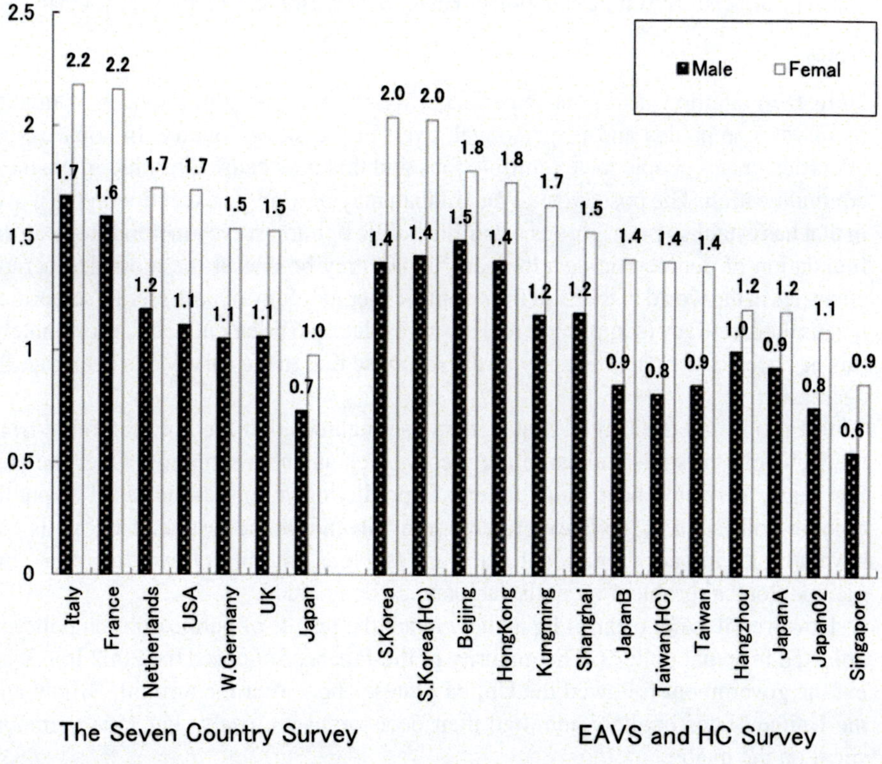

Fig. 3.10 Self-reported medical symptoms (the average number)

However, in most countries of the world, women are known to live longer than men. Moreover, the suicide rate of Japanese men is quite high in developed countries. These facts appear to be clearly inconsistent with the survey data above. All seven countries participating in this survey are highly industrialized countries. Therefore, there seems to be no significant difference in stressful situations or the quality of the healthcare system (The healthcare system may be linked to political issues in each country). On the other hand, women are said to be very good at relieving stress by chatting with other people, but men, especially Japanese men, tend to close their hearts and even commit suicide in the face of serious problems. Generally, the suicide rate is higher in men, but the suicide attempt rate is higher in women. Taken together, the above response pattern to medical symptoms appears to be closely associated with gender and country differences in self-disclosure.

The right side of Fig. 3.10 shows data from EAVS and a survey on health and culture (HC Survey) (Yamaoka, https://www.med.teikyo-u.ac.jp/~yama-hc/). This data may confirm gender differences as described.

When asked about life satisfaction, women may show higher *satisfaction* than men, but when asked about life *dissatisfaction*, women may show higher dissatisfaction. Such contradictions may be due to gender differences in self-disclosure and non-unidimensional structure of psychological satisfaction. In addition, the degree of self-disclosure may be closely related to the Japanese attitude of avoiding clear answers and extreme choices in polls (Yoshino, Hayashi & Yamaoka, [14, Chap. 3]).

Notes

Note 1. In addition, from our experience, we cannot ignore differences of quality in survey companies and in personnel, even in the same country. In some Asian countries, many people have multiple jobs and this may cause a trouble in business communication. The post-World War II Japan may be a single exception in the world in that has established a rigorous scientific public opinion survey methodology as the foundation of democratic development. Japan may be one of the most democratic countries in the world in terms of the scientific methodology of poll. Here, "scientific" means that accuracy (sampling error) of survey data can be calculated. Unfortunately, this has been questioned recently as the response rate to the survey has been rapidly decreasing.

In Europe and the United States, they use methods that are not preferable from the viewpoint of statistical sampling theory, such as quota sampling. Over the past few years, there have been many failures in predicting voting results based on public opinion polls such as the Brexit referendum. Given that the results of US polls can have a major impact on national decisions (such as entering or avoiding war), the methodological problem on polls can damage democracy.

However, it is not clear in Japan how much the results of public opinion polls are reflected in actual politics. The majority of the Japanese opposed the 2002 Iraq War, but the government followed the United States. Then, after the turmoil, Britain and the United States publicly admitted their decision as a mistake, but Japan remains silent on the matter.

Note 2. With these methods, usually, the sampling probability for each household is the same, but the sampling probability for each respondent is inversely proportional to the total number of adults (or eligible respondents) in each household. For a single adult household, he / she will always be sampled. If it is a household of four adults, for example, each of the four individuals will be sampled with a probability of 1/4. Therefore, to equalize the sampling probabilities at the individual level as a whole, the response data extracted from a household of four adults should be weighted four times as much as from a single household. However, this "weighting adjustment" assumes an effective cooperation rate of 100%. In reality, the cooperation rate of middle-aged housewives is generally high, but that of young adults is quite low. Therefore, our survey data tend to include more housewives and fewer young men. Then, if the above weighting adjustment is applied, a housewife who usually has a large family has a higher weighting than a young adult who lives alone, and the bias becomes even larger. Some people apply both "weighting adjustment according to the number of family members" and "weighting adjustment according to the census distribution." However, these adjustments are in the opposite direction, so the bias can be more complex than the original data.

Incidentally, in order to avoid this contradiction of weight adjustment, Chikio Hayashi (Kitada [54]) proposed a method called the "local accumulation method" and incorporated it into the opinion poll of nuclear power plants. Simply put, when visiting a house (for example, by random-route sampling), gradually build a list of residents and at regular intervals (such as 5 residents) select respondents from the list. His method solves the problem area sampling weight adjustment because each respondent has theoretically equal sampling probabilities. However, in reality, obtaining a high cooperation rate was not technically easy and the survey did not last long.

Note 3. Many Indians not only have multiple jobs at the same time, but also have holidays for each religion, so it may not always be easy to get in touch promptly for business.

Note 4. Incidentally, you can also introduce a chain of links for various survey modes, such as face-to-face, landline, mobile, computer-assisted personal interview [CAPI], web survey and so on. This is important because longitudinal surveys can inevitably change survey modes over decades due to security, cost, technology, or other reasons. This is also the case for cross-country surveys, because each country may use its own survey method. You can also think of a chain of question items that have the same meaning but translated into different languages, or a chain of wordings that are the same item but slightly different. However, other possible chains will not be discussed in detail here, as it is necessary to avoid overcomplication early in the research.

Note 5. Behind the idea of connecting charts is the concept of "meaningfulness" in abstract measurement theory developed in mathematical psychology ([55, Chap. 2] [40]). This is closely related to the Erlangen program proposed by the mathematician Klein, who captured diverse geometries such as Euclidean geometry or Riemannian geometry in a unified way on group theory. In the program, the subject of geometry is to study "invariants under an admissible transformation." The stronger the mathematical conditions for the allowed transformations, the richer the invariants. This is an aspect of complementarity described previously. In the social sciences, rigorous abstract measurement theory is too restrictive, so we need a more lenient view of scale and statistical indicators.

Note 6. We must be careful regarding the distinction of native Taiwanese (those born and raised in Taiwan) and Chinese mainlanders (those who emigrated to Taiwan during the post-World War II political turmoil and their descendants). Although the distinction is important in analyses, people in Taiwan are sensitive to certain political matters. It is based on self-reporting in the survey. In our Taiwan survey, there were fewer Chinese mainlanders than we would expect from census data. There are at least two possibilities: our sample might have been truly biased, or the mainlanders might have hidden their identities for political reasons. The latter possibility may lead to no statistical difference of response distribution between the two groups. Therefore, the aforementioned interpretation should be considered tentative (Here I'm not saying that the quality of the survey is low and unreliable. But I would like to emphasize that social survey data cannot avoid certain political biases, but such biases themselves provide a deeper side of reality).

Note 7. An experimental study, called the "Ohajiki Survey," reports specific relationships in response distribution between individuals and their groups [41] (Ohajiki [marble] is a toy for children in the shape of a small glass disk.) Give each respondent 5 Ohajiki [marbles] and ask them to answer each question with the degree of pros and cons, such as "I agree: I disagree = 3:2." Relatively, many Americans answered "5:0 or 4:1 (or 0:5 or 1:4)," but many Japanese answered "2:3" or "3:2." This means that when asking questions with two choices, "agree" or "disagree," the observed response may not be a complete representation of the distribution of the mind of each individual. However, we found interesting results when comparing the surveys on the same set of questionnaires in the Ohajiki survey and the two-choice survey ("Agree" and "Disagree"). That is, the marginal distribution was found to be consistent between the two surveys (The proportion of those who answered "5:0, 4:1 or 3:2" in the ratio of "Agree: Disagree" in the Ohajiki survey was equal to the proportion of those who answered "Agree" in the two-choice survey [42]. Although this experiment may not be immediately generalizable, but it may suggest something important between the individual mind and the group mind.

Another study also touched on individual-level variability and group-level stability. In an experimental panel survey, the same question was asked three times at regular intervals (weeks or months) on a sample of the same respondents. The question was about everyday life issues that seemed to remain unchanged for months. However, about 30–50% of respondents changed their answers to the same question on any of the three waves. On the other hand, for the entire group of respondents, the marginal distributions of "yes" and "no" have not changed much (up to a few percent change) [56, 43, 47]. This means that for no apparent reason, individual responses may change, but overall group responses may be stable.

3.5 Summary

Since 1971, Japanese National Character Survey (JNCS) has been expanded to include oversea surveys to better understand Japanese national character in the context of cross-national comparison. The survey items were selected to compare people's social values, ways of thinking, emotions, religious attitudes, etc. These aspects may provide information on the psychological distance between countries or races.

Cross-national survey needs to overcome the multifaceted methodological problems of cross-national comparability. These include problems on (1) translation (survey questions need to be created to maintain the same meaning in different languages), (2) comparison of datasets collected using different sampling procedures in different countries, and (3) descriptions and characteristics of the compared countries in terms of a common logic (or framework of thinking). Comparing people from different cultures makes these problems even more serious. Multidimensional data analysis, such as Hayashi's Quantification Method III (QMIII), often helped overcome translation problems and the use of different sampling methods in cross-national survey.

In a cross-national comparative survey, comparing completely different countries from the beginning is not the best way to make a meaningful comparison in our type of survey. By comparing pairs of countries (or groups) with similarities, such as language and ethnicity, and identifying similarities and differences, you can reveal more meaningful statistical comparisons. Gradually connecting these comparison links (country pairs) will expand the chain of links and ultimately allow for global comparisons.

This idea was developed as a research paradigm called "Cultural Link Analysis (CLA)" and eventually integrated three types of linkages: (1) longitudinal (temporal) linkage, (2) cross-national (spatial) linkage, (3) thematic linkage (item-structure linkage). We have also developed "Cultural Manifold Analysis (CULMAN)" to introduce a hierarchical structure into the three types of linkages. Under these research paradigms, our early research in cross-national survey revealed attitudes and social values particular to the Japanese, such as interpersonal relationships and religion. In addition, some survey results have been reported on the general response tendency and the degree of self-disclosure particular to each country (e.g., the Japanese tend to avoid extreme answers and choose a category near the middle of the options, or say "undecided" or "don't know").

The idea of CULMAN is expected to serve as a framework for developing empirical social sciences to understand the rise and fall of civilization and promote mutual understanding between different cultures.

References

1. Gill, C. J., & Gill, G. C. (2005). Nightingale in Scutari: Her legacy reexamined. *Clinical Infectious Diseases, 40*(12), 1799–1805.
2. Hayashi, C. (Ed.). (1973). Japanese Americans in Hawaii. *ISM Research Report*, 33. Tokyo: The Institute of Statistical Mathematics.
3. Hayashi, C. (1998a). What is data science? Fundamental concepts and a heuristics example. In C. Hayashi, K. Yajima, H. H. Bock, N. Ohsumi, Y. Tanaka, & Y. Baba. (Eds.), *Data science, **classification, and related methods, Proceedings of the Fifth Conference of the International Federation of Classification Societies (IFCS-96)***, Kobe, Japan, 27–30 March 1996 (pp. 40–51). Springer.
4. Hayashi, C. (2001b). *De-ta no kagaku* (Science of Data). Asakura-syoten: Tokyo.
5. Hayashi, C. (1998b). The quantitative study of national character: Interchronological and international perspectives. In M. Sasaki (Ed.), *Values and attitude across nations and time* (pp. 91–114). Boston: Brill.
6. Hayashi, C. (1993a). ***Quantification method***. Tokyo: Asakura shoten (in Japanese).
7. Hayashi, C., Yoshino, R., Suzuki, T., Hayashi, F., Kamano, S., Miyake, I., Murakami, M., & Sasaki, M. (1998). *Kokuminsei nanaka-koku hikaku* (cross-national comparison of seven nations). Tokyo: Idemitsu-syoten.
8. Yoshino, R., Hayashi, C., & Suzuki, T. (1995). Kokuminnsei no kokusai-hikaku-chousa no tame no situmon-bun no sakusei (Construction of a questionnaire for a cross-national comparative social survey). *Japanese Journal of Behaviormetrika, 22*(1), 62–79.
9. **Yoshino, R., Nikaido, K., & Ujiie, U. (eds.). (2012). The Asia-Pacific values survey—Cultural** Manifold Analysis (CULMAN) on people's sense of trust. *ISM Survey Research Report*, No. 105. Tokyo: The Institute of Statistical Mathematics. Retrieved from https://www.ism.ac.jp/editsec/kenripo/pdf/kenripo105.pdf.
10. ESS. (2018). *European Social Survey Round 9 sampling guidelines: Principles and implementation.* https://www.europeansocialsurvey.org/docs/round9/methods/ESS9_sampling_guidelines.pdf.
11. Suzuki, T. (1996). A study on cross-national sampling survey. *Japanese Journal of Behaviormetrics, 23*(1), 46–62.
12. De Leeuw, E., Hox, J., Silber, H., Struminskaya, B., & Vis, C. (2019). Development of an international survey attitude scale: Measurement equivalence, reliability, and predictive validity. *Measurement Instrument for the Social Sciences, 1*(9), 1–10. https://doi.org/10.1186/s42409-019-0012-x.
13. Guttmann, L. H. (1972). The concept of a common range: four applications anf four fallacies (unpublsihed paper), pp. 1–15. In S. Levy (Ed.). (1984). Lous *Guttman on theory and methodology: selected writings* (pp. 32–41). Aldershot, England: Dartmouth Publishing Company.
14. Yoshino, R., Hayashi, F., & Yamaoka, K. (2010). *Kokusai- hikaku deta no kaiseki* (Analysis of comparative survey data). Tokyo: Asakura-shoten.
15. Yoshino, R., & Khor, D. (1995). Complementary scaling for cross-national analyses of national character. *Behaviormetrika, 22*(2), 155–184.
16. Yoshino, R. (2005). East Asia values survey for the development of behaviormetric study of civilization on cultural manifold analysis (CULMAN). *Japanese Journal of Behaviormetrics, 32*(2), 133–146.
17. **Fujita, T., & Yoshino, R. (2009). Social values on international relationships in the Asia-Pacific** region. *Behaviormetrika, 36*(2), 148–165.
18. Yoshino, R. (2015). Trust of nations: Looking for more universal values for intrapersonal and international relationships. *Behaviormetrika, 42*(2), 131–166.
19. Inglehart, R. D. (1982, January). Changing values in Japan and the west. *Comparative Political Studies*, 445–479.
20. Inglehart, R. D. (2018). *Cultural evolution: People's motivations are changing, and reshaping the world.*. London: Cambridge University Press.

21. Benzecri, J.-P. (2019). *Correspondence analysis handbook.* Routledge.
22. Nishisato, S. (1994). *Elements of dual scaling: An introduction to practical data analysis.* Psychology Press.
23. Young, F. W., de Leeuw, J., & Takane, Y. (1976). Regression with qualitative and quantitative variables: An alternating least squares method with optimal scaling features. *Psychometrika, 41*, 505–529.
24. Yoshino, R., & Hayashi, C. (2002). An overview of cultural link analysis of national character. *Behaviormetrika, 29*, 125–142.
25. Yoshino, R. (1992). Superculture as a frame of reference for cross-national comparison of national characters. *Behaviormetrika, 19*(1), 23–41.
26. Yoshino, R. (2001b). *Kokoro wo hakaru* (Measuring the mind). Tokyo: Asakura-syoten.
27. Romney, A. K., Weller, S. C., & Batchelder, W. H. (1986). Culture as consensus: A theory of culture and informant accuracy. *American Anthropology, 88*(2), 313–338.
28. Batchelder, W. H., & Romney, A. K. (1988). Test theory without an answer key. *Psychometrika, 53*(1), 71–92.
29. Yamamoto, K., Kawai, T., Wakisaka, K., Miyao, S., Mori, K., Hayashi, C., Midzuno, H., Suzuki, T., Hayashi, F., & Yoshino, R. (1993). Research on national character of Japanese Brazilian—1991–1992. *ISM Research Report,* No. 74.
30. Narens, L. (1985). *Abstract measurement theory.* Boston: MIT Press.
31. Yoshino, R., Shibai, K., & Nikaido, K. (eds.). (2015). Sense of Trust—Summary Report: The Asia-Pacific Values Survey—Cultural Manifold Analysis (CULMAN) on People's. *ISM Survey Research Report*, No. 117.
32. Yoshino, R. (2009). Reconstruction of trust on a cultural manifold: Sense of trust in longitudinal and cross-national surveys on national character. *Behaviormetrika, 36*(2), 115–147.
33. Dogan, M. (2000). Deficit of confidence within European democracies. In M. Halller (Ed.), *The making of the European Union* (pp. 243–261). Oaris: Springer.
34. Inkeles, A. (1997). *National character.* . New Brunswick: Transaction Publishers.
35. Hayashi, C., & Suzuki, T. (1997). *Shakai-chosa to suryo-ka* (Social survey and quantification methods). Tokyo: Iwanami-Shoten.
36. Yoshino, R. (2014). Trust of nations on cultural manifold analysis (CULMAN): sense of trust in our longitudinal and cross-national surveys of national character. In M. Sasaki (Ed.), *Cross-national studies on sense of trust*, Chap. 7. Tokyo: Chuo-University Press.
37. Umehara, T. (2002). *Creation of universal human philosophy*. Umehara Takeshi Collection 17. Tokyo: Shogaku-kan (in Japanese).
38. Fugita, S. S., & Fernandez, M. (2004). *Altered lives, enduring community: Japanese Americans remember their World War II incarceration* (The Scott & Laurie Oki Series in Asian American Studies). Seattle: University of Washington Press.
39. Fugita, S. S., Kashima, T., & Miyamoto, F. S. (2002). Methodology of comparative studies of national character. *Behaviormetrika, 29*(2), 143–148.
40. Narens, L. (2002). *Theories of meaningfulness.* London: LEA.
41. Hayashi, C., et.al. (1984). Statistical methods for comparative study of social attitude by surveys. *ISM Research Report General Series*, No. 59. Retrieved from https://www.ism.ac.jp/editsec/kenripo/pdf/kenripo059.pdf.
42. Hayashi, C., Hayashi, F., & Akuto, H. (1981). A Statistical method for comparative study of fundamental structure of social attitude. *ISM Research Report General Series,* No. 54. Retrieved from https://www.ism.ac.jp/editsec/kenripo/pdf/kenripo054.pdf.
43. Hayashi et al. (1971). A statistical study on the response uncertainty in opinion survey. *ISM Research Report General Series*, No. 26. Retrieved from https://www.ism.ac.jp/editsec/kenripo/pdf/kenripo026.pdf.
44. Hayashi, C. (2001a). *Nihon-jin no kokuminsei kenkyu* (A study on the Japanese national character). Tokyo: Nansou-sya.
45. Hayashi, F., & Yamaoka, K. (2002). *Chosa no jissai* (Practice of surveys). Tokyo: Asakura-Syoten.

46. Midzuno, H., Hayashi, C., & Aoyama, H. (1953). ***Quantification and prediction—A study on*** *prediction of election outcomes.* Tokyo: Maruzen, Co. (in Japanese)
47. Suzuki, T., et al. (1981). Sources of measurement errors and their impact on survey research. *ISM Research Report General Series,* No. 52. Retrieved from https://www.ism.ac.jp/editsec/kenripo/pdf/kenripo052.pdf.
48. Suzuki, T., & Yanagihara, R. (2003). Census on Japanese language: Sampling methods and survey practice. *Japanese Journal of Behaviormetrics, 30*(1), 7–29.
49. Yoshino, R. (2012). Reconstruction of trust on a cultural manifold. In M. Sasaki & R. M. Marsh (Eds.), *Trust: Comparative Perspectives.* (pp. 297–346). Leiden: Brill Academic Publishers.
50. Yoshino, R. (2013). Trust of nations on cultural manifold analysis (CULMAN): Sense of trust in our longitudinal and cross-national surveys of national character. In N. I. Dryakhlov, A. Ishikawa, A. B. Kupreychenko, M. Sasaki, Z. T. Toshchenko, & V. D. Shadrikov (Eds.), *Trust in society, business and organization: Proceedings of the conference, "business, society, human"* (pp. 213–250). Moscow: National Research University.
51. Yoshino, R., Nikaido, K., & Fujita, T. (2009). Cultural manifold analysis (CULMAN) of national character: Paradigm of cross-national survey. *Behaviormetrika, 36*(2), 89–113.
52. **Yoshino, R., Shibai, K., Nikaido, K., & Fujita, T. (2015). The Asia-Pacific values survey 2010–** 2014—Cultural manifold analysis (CULMAN) of national character. *Behaviormetrika, 42*(2), 99–130.
53. Kuroda, Y., & Suzuki, T. (1991). Arab students and English: the role of implicit culture. *Behaviormetrika, 29,* 23–44.
54. Kitada, A. (2011). A study of two types of sampling methods for the INSS'07 opinion survey. *Japanese Journal ofBehaviormetrika, 84,* 13–32.
55. Yoshino, R., Chino, N., & Yamagishi, K. (2007). *Mathematical psychology (in Japanse.).* Tokyo: Baihuu-kan.
56. Takahashi, K., Nagasaka, K., Nakamura, T., Moroi, K., & Suzuki, T. (1980). Source of measurement errors and theirimpact on survey reseacrh. ISM Research Report, No. 49. The Institute of Statistical Mathematics.
57. Hayashi, C. (2000). The future of quantitative study on national chraracter - to quantitative study of civilaization fromcomparatuve study of national chracter (in Japanese). *The proceedings of ISM, 48*(1), 33–66.
58. **Osumi, N. (2003). In memory of Dr. Chikio Hayashi—from classification to data science (in Japanese). Japan Classification Society News, No.26.** http://bunrui.jp/pdf.kaiho26.pdf
59. Yoshino, R. (1992). Extension of the "test theory without answer key by Batchelder and Romney" and its application to an analysis of data of national consciousness. *Proceedings of the Institute of Statistical Mathematics, 37*(2), 171–188.
60. Yoshino, R. (1992). The unbiased BIGHT model and its application to the distinction of responses to a free answerquestion in a social survey. *Behaviormetrika, 19*(2), 83–96.
61. Yoshino, R. (1998). A social quantum theory for the analysis of public opinion survey data. *Behaviormetrika, 25*(2), 111–132.

Chapter 4
Development of CLA & CULMAN

4.1 Early Comparative Analysis of Japanese, Overseas Nikkei, and Westerners

Hayashi [1] reviews our longitudinal and cross-national surveys up to the early 1990s. His paper covers our cross-national surveys conducted in Japan, the United States and three European countries (UK, France, and West Germany), as well as overseas Nikkei (Hawaii and Brazilian Japanese emigrant societies) (see Table 3.1). Although it describes a wide range of topics, some of the key findings are summarized below.

4.1.1 "Giri-Ninjyo" Scale: Measurement of Japaneseness

First of all, the characteristics of Japanese people are recognized regarding the "preferred image of boss at work." The related question is phrased as below.

#5.6 Type of boss preferred.

(Hand card) Suppose you are working in a firm. Which of the following department chiefs would you prefer to work under?

1. A person who always sticks to the work rules and never demands any unreasonable work, but who, on the other hand, never does anything for you personally in matters not connected with work
2. A person who sometimes demands extra work in spite of rules against it, but who, on the other hand, looks after you personally in matters not connected with work

The second type of boss is nicknamed as "Ninjyo-Katyo (heartful and personal boss)." Overwhelmingly, the Japanese people prefer the "Ninjyo-Katyo." This tendency lasts stable and is stronger than in Europe and the United States. This is an example that basic interpersonal relationships remain stable long. (As explained

R. Yoshino, *Cultural Manifold Analysis on National Character*,
Behaviormetrics: Quantitative Approaches to Human Behavior 10,
https://doi.org/10.1007/978-981-16-1673-0_4

in Sect. 2 of Chap. 2, the "giri-ninjyo" attitude is also a typical Japanese value in interpersonal relationships).

The questions of preferred boss and of "giri-ninjyo" may be country-specific, so these may not be used for *direct* cross-national comparison without adequate modifications. (Each country has its own employment system and work style. Therefore, without proper modification, this type of questions may not be valid as the same in other countries). However, Hayashi uses the question items of "giri-ninjyo" and "Ninjyo Katyo" with some other items as several scales to measure Japaneseness in comparison with Japanese immigrants and Westerners. More concretely, the "giri-ninjyo" scale consists of questions #4.4(1), #5.1(1) & 5.1b(2), #5.1c1(1) & 5.1c2(2), #5.6(2), and #5.1d(1&2) from JNCS [1] (For Code# and wordings of questions such as #4.4, see [2], or https://www.ism.ac.jp/kokuminsei/en/table/index.htm). Here, for example, #4.4(1) means the scale assigns 1 point to the choice "1" of questions #4.4. And #5.1d(1&2) means the scale assigns 1 point to the simultaneous choice of "1" and "2" of question #5.1d (multiple choice question). It makes a 5-point scale. The data of JNCS confirms the stability of distribution of Japanese on this index over decades ([3], Table 2.5). Figure 4.1a shows data to compare Japanese, Americans, and Europeans. It is clear that Japanese are different from the others. Furthermore, Fig. 4.1b compares Japanese, Japanese-Americans, and Japanese-Brazilians.

The "ninjyo" scale uses the same items but assigns 1 point as #4.4(1), #5.1(1), #5.1b(1), #5.1c1(2), #5.1c2(2), #5.6(2), #5.1d(1), and #5.1d(2). This is an 8-point scale. See Table 4.1.

On these scales of "giri-ninjyo" and "ninjyo," Nikkei (Japanese immigrants) are generally closer to Japanese in terms of relationships. However, Hawaiian Nikkei are leaning toward the United States for preference taste of the workplace boss (not the "Ninjyo Katyo"). Brazilian Nikkei don't necessarily want to work with close friends. But overall, Hawaiian-Japanese and Brazilian-Japanese are generally similar, although Japanese-Brazilians are more like Japanese. These observations may be related to the differences between those Japanese communities in terms of their social distance to other ethnic communities. Although the multiethnic populations coexist on the smaller islands of Hawaii, the Japanese-Brazilians may inevitably tend to live closer together to help each other in the much more huge lands of Brazil. But even those Japanese immigrant societies may change over time or with generational changes. (Later analysis by Yoshino [4], including data from JAWCS [the Japanese Americans on the West Coast Survey), suggests that the comparison of Japanese migrants was more complex, and their similarities or differences differed from issue to issue).

4.1.2 Tradition Versus Modernization

Hayashi [1] presents also some analysis to investigate the Japanese way of thinking on tradition versus modernization since the start of Meiji restoration era (1868, i.e., the end of the Samurai era). For the longitudinal survey data of JNCS, he carried out

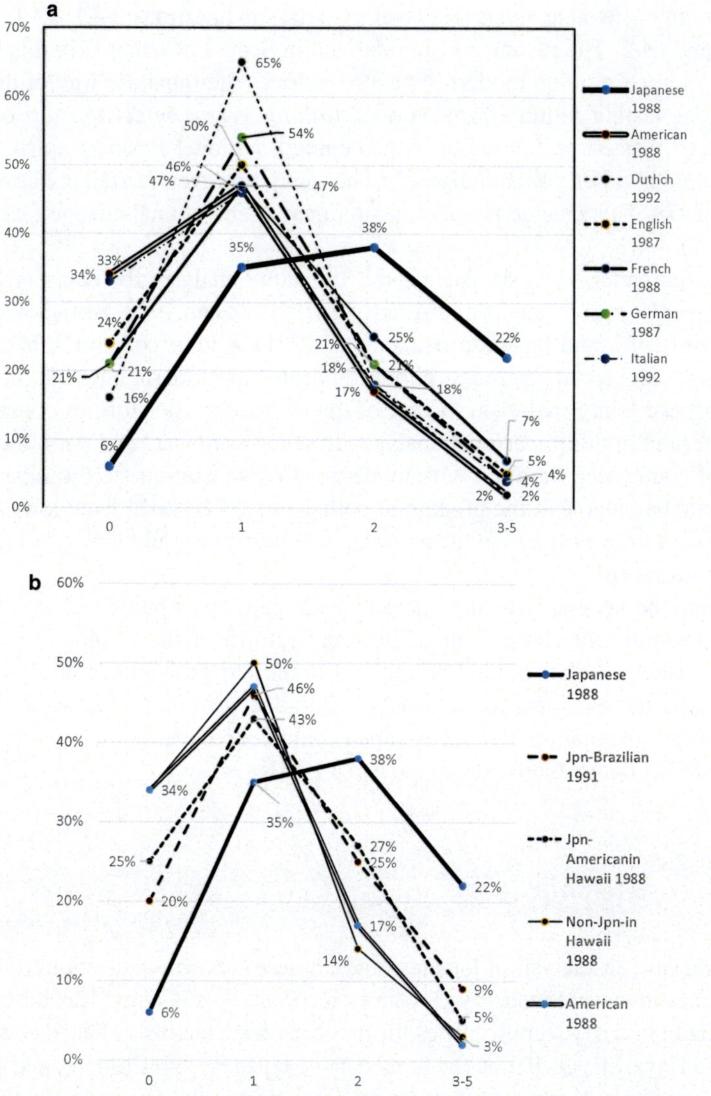

Fig. 4.1 **a** Distribution of Giri-Ninjyo Value Index (0–5 point). There is a clear difference between Japanese and Americans or Europeans in this index. (Data from Hayashi and Kuroda [3], Table 3.4). **b** Distribution of Giri-Ninjyo Value Index (0–5 point) Japanese Americans in Hawaii and Japasene Brazilians are somewhere between Japanese and Americans on this index. (Data from Hayashi and Kuroda [3], Table 3.4)

Table 4.1 Ninjyo Value Index (0–8 point) Percentage (%) of people with 5 points or more

Japan	French	Japanese Brazilian	German	Japanese Americans in Hawaii	English	Americans	Non-Japanese Americans in Hawaii
38	35	34	33	29	27	23	19

a multidimensional analysis (Hayashi's QMIII) on questions: #4.1, #2.1, #2.5, #6.4, #8.1, and #4.5. The results are summarized in Sect. 2 of Chap. 2 in this book, with some interpretation on modern Japanese history. The Japanese way of thinking has been long-lasting within a framework of *tradition versus modernization*, or acquiring advanced Western technologies with keeping traditional Japanese spirit. (This kind of attitude is called "*Wakon-Yosai.*") The trend, however, started collapsing around 1973–1978. This change is due to a significant generational change (see Sect. 2 of Chap. 2).

In this matter, Hayashi carried out the same analysis for datasets from other countries (Hawaii residents and Americans of 1978 and 1988, British, German, and French of 1987, and Japanese-Brazilian of 1991). And he concludes that each question item can be categorized as either traditional or modern, but with the exception of Japanese living in Japan, the set of questions does not form a consistent scale or pattern in multidimensional analysis. In other words, at least until the 1990s, the idea of contrasting tradition with modernization was unique to the Japanese living in Japan, but is now in the process of collapsing. (If Hayashi lived long and looked at our data from East or Southeast Asian countries, he could have found a situation similar to Japan).

It may be necessary to note here that his paper was published in 1993, before Japan's significant change under the rapid reform of the economic and political system called "globalization." In the 1990s, the Japanese witnessed the collapse of even basic interpersonal relationships at home, at school, and at work. This seems to be one of the main reasons why Japan could not recover in the long term after the "Bubble Economic Collapse" in the early 1990s.

4.1.3 Leadership

Hayashi gave an analysis of Japanese and Chinese surveys on questions of leadership. (The questions were made by G. Chu of the East–West Center, Hawaii University). He did not find any significant results in the separate analysis of each country. So he applied Hayashi's QMIII on the bond data of Japanese and Chinese, and identified a clear distinction of preference of leadership between the Japanese and the Chinese. (The discrimination rate is 88% on the score of the first dimension of Hayashi's QMIII). The distinction is summarized in Table 4.2.

It should be noted that his analysis is based on data from the early 1990s. The last three decades have seen rapid changes in both Japan and China. Some political and economic changes may have brought about significant changes in social values, including interpersonal attitudes, in each country. Then came a new image of political and corporate leaders. However, the basic interpersonal attitudes are important for long-term national stability. In the name of globalization, even basic interpersonal relationships were endangered, jeopardizing social stability. Over the last three decades, Japan has seen a strong shift toward Americanization of basic interpersonal relationships, including leadership, in the 1990s, but is returning to

Table 4.2 Features of Japanese Leadership and Chinese Leadership

Japanese leadership	Chinese leadership
• Respected and liked by subordinates	• Bring benefits to his/her subordinates
• Treat friends with sincerity	• Technically superior
• Good relationships and wide faces	• Young and talented
• Experienced and respected elderly people	• Decisive and determined
• Seniority	• Work seriously
• Good judgment	(Moral Ethics)
• Treat subordinates fairly	• Parental mercy and child filial piety
(Moral Ethics)	• Justice and morality
• Harmony and moderation	
• Tolerance and courtesy	

tradition after experiencing the "lost 20 years" of the economy. This reminds us that Japanese history repeats a pendulum-like process that fluctuates between tradition and modernization (foreignization) for hundreds or even thousands of years.

4.1.4 Science and Heart/Mind

As a view of scientific civilization, Hayashi uses several questions related to science and daily life as follows. Code "#" is for the Japan National Character Survey (JNCS) and "Q" for the Seven-Country Survey (SCS), the Pacific Value Survey (PRVS), or the Asia Pacific Value Survey (APVS). For example, #7.36 is from JNCS, and PRVS Q25 is from PRVS question 25. (For codes and wording of questions, see ISM survey reports: No. 91, No. 95, No. 119 from https://www.ism.ac.jp/editsec/kenripo/contents_e.html, or https://www.ism.ac.jp/~yoshino/index_e.html).

- Utilization of Science in Daily Life (#7.36)
- Information Technology for Daily Life (#7.33, also related to #7.1 and #7.2, Q25 of PRVS, Q32 of APVS)
- Views on Modern Science (#7.83, #7.84, #7.85; Q41a,b,c,d of APVS): (a) Western versus Non-Western Medical Treatment, (b) Complete Analysis of Deep Human Mind, and (c) Complete Solution of Economic and Social Issues.
- Q58a,b,c of SCS, & Q41d of APVS (Degrees of Possibilities of Science within 25 years: (a) Safe Nuclear Power, (b) Complete Treatment of Cancer, (c) Complete Treatment of Senile Dementia, and (d) Living on a Space Station.

Table 4.3 shows the response data for Japan, the United States, and five European countries (I added data of Italy and the Netherlands to Table 15 of Hayashi [1]). In general, the Japanese are positive about the usefulness of science. However, the

Table 4.3 Values on Science in the Seven Country Survey (1987–1993)

(Positive %)		Japan (1988)	West Germany (1987)	France (1987)	UK (1987)	USA (1988)	Italy (1992)	NDL (1993)
• Science is useful in daily life	"A lot"	48	37	33	50	61	53	22
• Development of PC is desirable	"Desirable"	31	15	32	16	34	48	19
• Non-western medical treatment may be better for some diseases	"Strongly agree" & "agree"	72	77	62	76	72	49	70
• Science eventually can elucidate human minds completely		14	34	65	50	58	52	32
• Scientific technology eventuay solve almost all social problems		15	44	49	43	47	53	35
• Safe method of nuclear power waste disposal	"Very likely" & "possible" within next 25 years	33	17	28	34	32	82	64
• Complete cure for all cancers		65	37	68	64	62	94	88
• Complete treatment of senile dementia		29	17	23	12	25	64	79
• Ordinary people's lives in space		19	36	23	33	41	40	50

Japanese are negative in solving social problems and completely elucidating the human mind through scientific approaches. This is the characteristic of Japanese scientific views, as opposed to Westerners. Germans (West Germany in 1988) are relatively negative about science, but not as negative as Japanese about the usefulness of scientific approaches to social problems and the human mind. (This may remind us that Marx's theory and Freud's psychoanalysis appeared in the German cultural sphere).

4.1.5 Nature

A question on nature was mentioned in Sect. 2.4.4 of Chap. 2 (Fig. 2.5).

#2.5 Man and nature.

(Hand card) Here are three opinions about man and nature. Which one of these do you think is closest to the truth?

1. In order to be happy, man must follow nature
2. In order to be happy, man must make use of nature
3. In order to be happy, man must conquer nature

In order to consider more about "man and nature," let's consider two questions on scientific civilization from JNCS as follows:

#7.1 Science and loss of human feeling.

Some people say that with the development of science and technology, life becomes more convenient, but at the same time a lot of human feeling is lost. Do you agree with this opinion, or do you disagree?

1. Agree 2. Undecided/it depends 3. Disagree

#7.2 Mechanization and human feeling.

Some people say that no matter how mechanized the world gets, nothing can reduce the richness of human feelings. Do you agree with this opinion, or do you disagree?

1. Disagree 2. Undecided/it depends 3. Agree.

The longitudinal data of JNCS shows the increasing response of loss of human feeling over years up to 1973 since 1953, but after 1973 the trends look fluctuating. These questions are closely linked with the Japanese unique way of thinking on modern versus traditional, as described in Sect. 4.1.3. As of 1953, people who respected tradition were more conservative to scientific civilization than those who had a modern mindset. In 1973, however, those who had a modern way of thinking became more critical on scientific civilization. The change was significant among the young cohort of 20–29 years old at the point of 1953; they have changed their opinions when they became 40–49 years old. As for those who were 30 years old or older at the point of 1953, they generally kept the same opinions over years, so their modern way of thinking stayed optimistic on scientific civilization.

These observations motivated Hayashi to investigate whether, at point of 1953, those who claimed that "man must follow nature" were conservative and those who claimed that "man must follow nature" were innovative. Hayashi carried out comparative survey of Tokyo metropolitan city (more than 12 million residents) and Freiburg (a local city of Germany, 300,000 residents) concerning questions as to where one would like to tour and whether one like to take a walk in woods [5]).

Japan and Germany were defeated in World War II. However, the two countries have steadily recovered their economic power after the war and reached the world's top class in the 1980s, so they are often compared. It is often pointed out that Japanese and Germans adhere to the rules and punctuality. Our survey of seven countries (1987–1993) found that Japan and Germany have some similarities in interpersonal relationships, but some differences, for example, in family values (Hayashi et al. [6, Part II, Chap. 1). High preference of people of Tokyo to visit was "a mountain with a good view (23%)" and "a quiet lake (22%)," but "deep woods" were only 3%. Preference of the majority of Freiburg was "deep woods (55%)." In addition, as for these who like to take a walk in woods, Freiburg was 97% whereas Tokyo was 62%. In this way, there are considerable differences in the closeness to nature and forests between Tokyo and Freiburg, partly because of the difference in natural environment.

As for the question about the mystery of nature "Do you feel mystery in deep forests?," the majority of Tokyo (53%) is positive, but far greater in Freiburg (93%). In addition, there are two questions asked. "Should we intervene in the forest to keep it beautiful?" and "Do you like human-grown or pristine nature with a mix of farms and forests?" The answers in Freiburg to these two questions was clear. "Needs human intervention" and "nature raised by humans." It may represent the feelings of citizens of Freiburg with beautiful artificial nature in the background of the city.

On the other hand, in Tokyo, there was a contradictory opinion that some people prefer forests that have been maintained by human intervention, even though they believe that forests are beautiful without human intervention. The people of Tokyo were divided into three groups of approximately the same size. A naturalist group (like Freiburg residents), a realist group, and a "romantic" groups (those who believe that human intervention is needed to keep nature beautiful, but prefer nature without intervention, or vice versa).

In addition, some photos of forests with and without human intervention were presented and asked which landscape was preferred. Taken together, these results favor a beautiful forest with human intervention (clean forest, beautiful forest). It turns out that there is no big difference in preferences among the previously classified three groups. In other words, there is little relation between ideological preferences for nature and perceived preferences. There seems confusion in the Japanese preference for Japanese gardens (created "naturally" by humans) but seeing them beautiful as if it were totally natural without human intervention. There seems to be no relation between conceptual preference for nature and perceived preference.

Including some quantification data analysis on nature and religious feelings, Hayashi concludes that simple religious feelings and the pros and cons of human intervention for nature can be considered to be conceptually completely independent. It became clear that Japanese preferences for nature depend on what is beautiful and well organized (artificial or natural) and not on whether human intervention is necessary. This is related to the independence between Japanese preference for nature found in the literature and their current perspective on environmental issues. This discrepancy may lead to the misfortune of destroying nature. Therefore, in Japan, it is impossible to solve environmental problems solely by the ideas as nature lovers.

At the same time, by considering a rational ecosystem in an ecological sense, it is necessary to develop an aesthetically pleasing system and gain public consensus. (For EAVS data analysis, see [7]).

For the development of the CLA paradigm, see Yamaoka and Hayashi [8], who presented an analysis of cultural connections between Japanese, Japanese-Americans, and Americans. They are focusing on the topic of interpersonal attitude, religious attitude, and way of thinking.

Incidentally, Japanese-American researchers, Miyamoto, Fugita, and Kashima [9] developed a theory of interpersonal relations, which was derived from G. H. Mead's approach. The theory focuses on people's balancing between direct behavior and deep cognition with the purpose of facilitating an understanding of the differences in character between Japanese and Americans. We collaborated on a West Coast survey (Seattle and Santa Clara) for Japanese-Americans. They indicated that the Japanese prefer delayed action and pay more attention to the perceptions of others, of themselves, and of the situation, whereas Americans favor direct action and pay more attention to their own views. Their theory was originally developed for the study of Americans with Japanese ancestry and their transition of generations from first ("Issei") to second generation ("Nisei"), and then from second to third generation ("Sansei"), and so on. This theory may lead to an understanding of the interaction between racial origins and social environments. (See also Kashima et al. [9]; Fugita et al. [10], Fugita et al. [11]).

Working with Miyamoto's team, I realized that Japanese-Americans have a dual thinking framework (mindset) of Americans and Japanese-Americans in their daily lives. For example, suppose they're asked if they respect their ancestors over the average person (cf. Q51 of JAWCS & Q5 of APVS) [12]. In answer to this question, they point out that they first wonder whether "the average person" means "the average American" or "the average Japanese-American." Furthermore, even with the same question, it may be necessary to consider different thinking frameworks also for different generations. Thus, in general, a minority group of people may have more or less multiple thinking frameworks in their daily lives.

4.2 Interpersonal Trust, Institutional Trust, and Other Social Values

4.2.1 Trust and People

During the Cold War, Rotter [13, p. 443] stated: "It seems clear that disarmament will not proceed without an increase in trust on one or both sides of the iron curtain." The iron curtain was torn down more than three decades ago, but new local conflicts have been occurring incessantly all over the world. Mutual understanding and mutual respect among countries are important for the peaceful development and economic prosperity of the world. And "trust" is the key to mutual understanding and respect.

This section reviews past research on people's sense of trust as reflected in our longitudinal and international data. Among other things, I focus on the trust system to investigate which aspects of trust are stable and which aspects fluctuate in the long term depending on economic or political circumstances. This can ultimately lead to the search for more universal social values that transcend national and time differences. (For a review of research on trust, see [14], Sect. 2 of [15]).

Some researchers say that "trust" cannot be measured directly. Probably there may be no universal measure of trust across cultures and times. Even if there is one, it may not necessarily be linear with respect to various factors such as income and social class. (Any food will harm your health, if overeaten). However, by properly analyzing the longitudinal patterns and cross-national patterns of questionnaire responses, it is possible to identify certain important aspects of people's sense of trust. Here it is important to consider the data, taking into account each country's social situation and the general response tendency (Sect. 3.4 of Chap. 3), rather than comparing the data superficially.

In the following, first, a summary is given on the Japanese sense of interpersonal trust as identified in JNCS. Secondly, a cross-national comparative analysis of interpersonal trust, institutional trust, and some universal values is explored.

4.2.2 Interpersonal Trust of the Japanese

The past decades have developed psychological studies of measures of interpersonal trust. Among others, a set of three question items from the General Social Survey (GSS) has been used to measure people's sense of trust (For some history, see [14, 16]). Although GSS started as a sort of American version of JNCS, we have adopted the three items from GSS for our survey since 1978. They are stated as follows (For the Japanese questionnaire, see https://www.ism.ac.jp/kokuminsei/index.html. See also Note 1 and Note 2.):

Q36. Would you say that, most of the time, people try to be helpful, or that they are mostly just looking out for themselves? 1. Try to be helpful, 2. Look out for themselves

Q37. Do you think that most people would try to take advantage of you if they got the chance, or would they try to be fair? 1. Take advantage, 2. Try to be fair,

Q38. Generally speaking, would you say that most people can be trusted or that you can't be too careful in dealing with people? 1. Can be trusted, 2. Can't be too careful.

These three questions are originally from five questions (trustworthiness, honesty, goodness, generosity, and brotherliness) of "Faith-in-People Scale" obtained by Rosenberg [17] by constructing the Guttmann scale with a 92% regeneration rate in a student survey. The five questions are also used for research such as Almond & Verba [18]. Later, three of the five questions were used in the ISR survey at the University of Michigan Survey Research Center and GSS at the University of Chicago NORC. The expression has been modified to become binary and is now often used in the above-mentioned form [14, 19].

The original five questions were extracted from many questions as independent factors. In particular, the above three questions are used as the scale of "trust in people." Question a is treated as a measure of "norm of reciprocity," b as "fairness," and c as a measure of "generalized interpersonal trust." However, we should pay attention to the labeling. The tendency to use them, as the labels indicate, is common in top-down studies that seek to prove a particular theory or model using empirical data, whether in politics, economics, or medicine [20, 21]. However, it should be noted that in many cases a single item (question) is not enough to capture one important potential factor. The number of items used on the scale and the statistical reliability of the scale are closely related. This is a kind of complementarity between the simplicity and reliability of scaling (cf. Section 3.2.1 of Chap. 3). Moreover, it is not obvious that a variable (question) has the same meaning in different countries.

As for question c, it has been reported that if the latter part of the expression is changed as "Do you think most people can be trusted, or do you think they cannot be trusted," the response rate of "can be trusted" increases in both Japan and the United States [22]. The expression of c is not asking "whether it can be trusted or not," but it is comparing two social values that do not necessarily contradict "whether you can trust" and "you should always be careful." We would presume some effect of "social desirability."

As mentioned, each of the three questions is supposed to capture somewhat different aspects of trust: Q36 "norm of reciprocity," Q37 "fairness," and Q38 "generalized interpersonal trust." It would be meaningful cross-nationally to compare the relative patterns of correlations between the three items country by country. See Table 4.4a–d. Most countries show roughly stable patterns (e.g., the pairs of highest correlations) over time, but some countries, such as South Korea, show variability. The variability might be caused by a slight change in wording or sampling methods. Our data repeatedly demonstrates that Q37 and Q38 are more correlated for the Japanese, whereas Q36 and Q37 are more correlated for Americans. Some countries such as India may consistently show unique patterns of correlations, whereas other countries such as South Korea may show less-stable patterns over years.

As a measure on a sort of "total interpersonal trust," I often use the percentage of those who choose positive categories to all of Q36, Q37, and Q38. (Because Q36 was missing in some of our past surveys, I sometimes use also the percentage of those who gave positive answers to both Q36 and Q38. The rankings on the two questions and on the three questions are mostly consistent over the countries/regions where the three questions were used. See Fig. 4.3 in the next section).

On the measure of "total interpersonal trust," both the Japanese and Americans have been fairly stable but the Japanese may be more stable than Americans, at least during 1983–2008 (Fig. 4.2a, b). We need to pay attention to the changes in 1978 and in 2013. The Japanese experienced a nationwide panic trying to hoard necessities for their daily lives after the oil crisis and the Nixon shock (i.e., the unilateral cancellation of the direct international convertibility of the US dollar to gold) around 1973. Necessarily, this would have downgraded mutual trust. On the other hand, in the disaster of the Great East Japan Earthquake of 2011, people run to the devastated area from all over Japan to help the suffered people. Still under the

Table 4.4 **a**. Asia-Pacific Values Survey (APVS): Spearman's Correlations between Pairs of Items Q36, Q37, and Q38. **b**. Pacific-Rim Values Survey (PRVS): Spearman's Correlations between Pairs of Items Q36, Q37, and Q38. **c.** East Asia Values Survey (EAVS): Spearman's Correlations between Pairs of Items Q36, Q37, and Q38. **d**. Seven Country Survey: Spearman's Correlations between Pairs of Items Q36, Q37, and Q38

a (APVS)

Year	2010	2010	2011	2011	2011	2011	2011	2012	2012	2013	2013
	Japan	USA	Beijing	Shanghai	Hong Kong	Taiwan	South Korea	Singapore	Australia	India	Vietnam
Q36 × Q37	0.14	**0.49**	0.19	**0.26**	**0.31**	0.28	0.00	**0.31**	**0.43**	0.04	0.10
Q36 × Q39	0.24	0.44	0.21	0.21	0.28	0.27	**0.23**	**0.29**	0.36	**0.25**	0.21
Q37 × Q39	**0.28**	0.39	0.21	0. 24	0.26	0.22	0.02	**0.32**	0.35	0.01	**0.35**

b (PRVS)

Year	2004	2006	2005	2005	2005	2006	2006	2007	2007	2008
	Japan	USA	Beijing	Shanghai	Hong Kong	Taiwan	South Korea	Singapore	Australia	India
Q36 × Q37	0.20	**0.53**	0.21	**0.24**	0.18	0.20	0.10	**0.32**	**0.44**	0.05
Q36 × Q39	0.30	0.36	**0.23**	0.21	0.19	**0.22**	0.24	**0.32**	0.38	**0.35**
Q37 × Q39	**0.31**	0.37	0.15	0.22	**0.22**	0.12	**0.26**	0.21	0.32	0.00

c (EAVS)

Year	2002	2002	2002	2002	2003	2003	2004	2003	2002
	Japan	Beijing	Shanghai	Hong Kong	Taiwan	South Korea	Singapore	Kunming	Hangzhou
Q36 × Q37	0.23	0.21	**0.33**	0.18	**0.26**	0.05	**0.32**	**0.37**	**0.27**
Q36 × Q39	0.32	**0.28**	**0.33**	0.19	**0.26**	0.13	0.31	**0.33**	0.21
Q37 × Q39	**0.37**	0.13	0.24	0.16	0.15	**0.24**	0.22	0.31	0.25

(continued)

Table 4.4 (continued)

d (Seven Country Survey)

Year	1988	1988	1987	1987	1992	1993	1987
Country	Japan	USA	France	UK	Italy	Netherlands	West Germany
Q36 × Q37	0.14	**0.45**	0.32	**0.44**	**0.40**	**0.36**	0.36
Q36 × Q39	0.24	0.37	0.21	0.35	0.30	0.29	**0.49**
Q37 × Q39	**0.29**	0.44	**0.38**	0.36	0.34	0.32	0.34

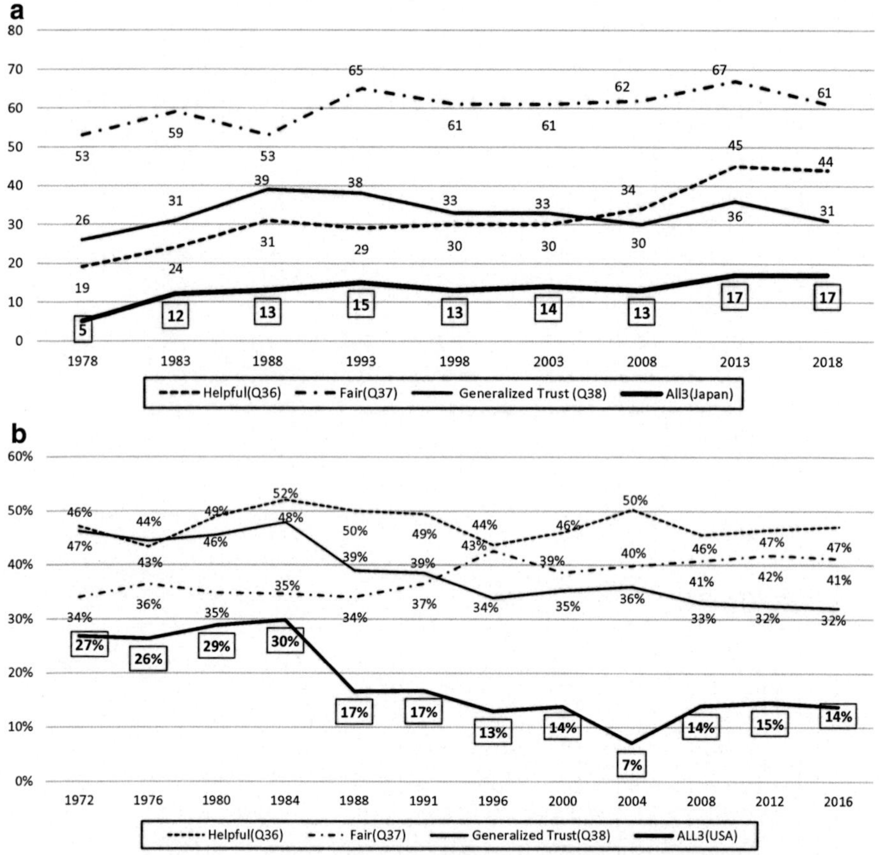

Fig. 4.2 **a** Percentages of Positive Responses of GSS Trust items (Q36, Q37 & Q38) and Those Who Gave Positive Responses to All the 3 Items in Japan. The Japanese data are from the Japanese National Character Survey, execept the1988 data from the Seven Country Survey. **b** Percentages of Positive Responses of Trust (Q36, Q37 & Q38) and Those Who Gave Positive Responses to All of the 3 Items in the USA.The data are from the SPSS data of GSS (downloded from the website of NORC, May 23, 2019)

lasting economic depression, many people considered how they could contribute to recovery of the area. Comparing to those suffered in the great disaster of earthquake and the succeeding nuclear power plant disasters, all the Japanese must have felt that they must be satisfied with their lives and have to appreciate mutual assistances. Naturally, this would have raised mutual trust.

If Q36, Q37, and Q38 are *separately* studied, the Japanese data also show more changes over the years. The changes may confirm that the economic and political structural reformation damaged the Japanese sense of trust, roughly, during 1993–2008.

After World War II, Japan's lifelong employment system provided better employment security. Workers' salaries may be lower during recessions, but they were not easily dismissed. This may explain the higher levels of trust (i.e., the relative stability of response patterns on Q36, 37, and 38) found in the Japanese than in the Americans, at least during 1983–2008. But, in the early 1990s, the reformation under "globalization" started to force the Japanese to change economic, political, and social systems, looking for efficiency or internationalization disregarding of the Japanese structures rooted in historical background and culture. The change of the social systems attacked even interpersonal systems of family, school, and workplace, disturbing people's heart and mind. The "lost two decades" since the collapse of bubbling economy around 1991, after all, resulted in confusions and failures not only in the Japanese systems but also in foreign banks and commercial companies which attempted to take advantage of the opportunities in Japan.

Meanwhile, the government has lost people's trust in the national pension system. Senior people rely on younger people for future financial support, but the population of younger generations has been decreasing, and the younger are less motivated to pay pension costs, in consideration of the balance. These situations, originally due to distrust on governmental institution, have necessarily led to a gap of consciousness on social institutions between the young and the senior people.

Incidentally, the new graduates during, roughly, 1993–2004 are called "Syusyoku Hyoga-ki Sedai" (Ice Age Generation of Job Market) or "lost generation" because they faced remarkable difficulties to get regular job positions under the rapid recession (Note 3). They are now in their 30 s or 40 s, but still face difficulties of getting positions of regular employees, in spite of recent economic recovery. Because Japanese job market is mainly for new graduates, those of Ice Age Generation meet much more disadvantages than younger graduates. At last the government started amending the situation, demanding the Japanese business world to employ them as regular employees.

Looking back over the last 40 years, during the prosperity of the 1980s, there was a shift of young men's social values toward individualism and then personal preference (give priority on personal matters). But the structural reforms of the 1990s led to the economic recession and the departure from lifetime employment. As a reaction, human relationships in the workplace seem reconsidered (regression to tradition), among others, in young people.

Yoshino [22] discussed several aspects of trust, such as trust in politics, science, and technology, as well as the work ethic of the Japanese, and concluded that some aspects of trust may be variable according to economic and political conditions, whereas some other may be more stable. For example, although the world used to have a stereotype of the hard-working Japanese called the "economic animal" in the 1980s, their attitudes and ethic toward work seem to be influenced by economic and political conditions. (See Inagaki and Maeda [23] for the study of trust using latent structure modeling on JNCS).

(As for an important comment for the relationships between survey cooperation rate and respondent's sense of trust, see Sect. 5.7 of Chap. 5).

4.2.3 Cross-National Survey on Trust and Universal Social Values

This section outlines some studies of trust in our past cross-national surveys, covering the topics of interpersonal trust and institutional trust. It includes a survey of Japanese immigrants and various ethnic groups in some countries. This transcends national differences and explores the universal value underlying the trust of people.

4.2.3.1 Sense of Interpersonal Trust

Our cross-national surveys also included the three questions on interpersonal trust from GSS. Table 4.5a-g shows the data from most of the countries/regions that we have surveyed over the past four decades. See Fig. 4.3 also.

Miyake (Hayashi et al. [6], Chap. 7) presented an analysis on our Seven-Country Survey: Japan, the United States, UK, West Germany, and France in 1987–1988 and Italy and the Netherlands in 1992–1993. He concluded that the trust scale had correlations with gender and religion but stronger correlations with family income, educational level, and social class. (But these correlations were not so strong). On this scale, West Germany, UK, and the United States scored higher than Japan and the Netherlands, but the difference was small. The French and Italians clearly scored lower than in other countries. In addition, those who had religious faith gave more positive response rate ("try to be helpful") to Q36, irrespective of their religious affiliation. For Q37, women gave more optimistic answers ("they would try to be fair") than men. As for Q38, there was a clear difference between social classes in all seven countries. That is, the higher the social class, the more trustful the respondents were. The difference between classes was remarkably large in France and the United States. Observing that the higher level of education was associated with the greater trust, Miyake reasoned that the association was caused by the correlation between education level and social class. (Although there was a relatively strong correlation between education level and social class, and between education level and income in the United States, this was not necessarily the case for other countries).

Using the same data, Yoshino [22] showed positive correlations between trust and social class or income level in the United States and UK but nonlinear correlations in the other five countries, including Japan and West Germany (i.e., the middle social class shows the higher trust rate than the lower and the higher class). Incidentally, Yoshino & Tsunoda [24] suggested nonlinear relationship between subjective health and sense of trust.

Figure 4.3 shows the ranking of the percentages in each country of those who gave positive answers to both Q36 and Q38. The measure seems fairly stable within the countries/regions when repeatedly surveyed over years. Interestingly, the Japanese immigrants in the United States and Brazil are ranked, respectively, as the highest and the lowest. That is, Japanese-Americans on the West Coast (JAWCS) is higher than the general Americans, whereas JBs (Brazilians of Japanese descent) are lower than

Table 4.5 **a.** Percentages (%) of Positive Responses to GSS 3 Items on Trust in the Asia-Pacific Values Survey (APVS). **b.** Percentages (%) of Positive Responses to GSS 3 Items on Trust in the Pacific-Rim Values Survey (PRVS). **c.** Percentages (%) of Positive Responses to GSS 3 Items on Trust in the East Asia Values Survey (EAVS). **d.** Percentages (%) of Positive Responses to GSS 3 Items on Trust in the Japanese National Character Survey (JNCS). **e.** Percentages (%) of Positive Responses to GSS 3 Items on Trust in Seven Country Survey. **f.** Percentages (%) of Positive Response to GSS 3 Items on Trust: Breakdown by Ethnicity for Singapore Surveys in 2004, 2007 & 2012. **g.** Percentages (%) of Positive Response to GSS 3 Items on Trust: Comparison of Japanese Immigrant Surveys

a (APVS)

Q36	People are always trying to be helpful to others
Q37	People are trying to be fair
Q38	People can be trusted

Year	2011	2011	2011	2011	2012	2012	2013	2012	2010	2010	2010
APVS	Beijing	Shanghai	Hong Kong	Taiwan	South Korea	Singapore	India	Australia	USA	Japan	Vietnam
Q36	72	66	43	46	52	50	55	59	51	41	70
Q37	57	58	40	53	53	49	34	63	53	57	21
Q38	42	36	21	21	32	34	45	45	31	44	23

b (PRVS)

Survey Year	2005	2005	2005	2006	2006	2007	2008	2007	2006	2004	2004
PRVS	Beijing	Shanghai	Hong Kong	Taiwan	South Korea	Singapore	India	Australia	USA	Japan A	Japan B
Q36	67	65	41	46	57	50	61	54	56	35	37
Q37	51	53	42	61	45	51	29	59	56	59	–
Q38	37	34	19	19	30	26	52	43	41	37	39

(continued)

Table 4.5 (continued)

c (EAVS)

Survey Year	2002	2002	2003	2002	2002	2003	2003	2004
EAVS	Beijing	Shanghai	Kunming	Hangzhou	Hong Kong	Taiwan	South Korea	Singapore
Q36	62	59	48	61	36	38	58	51
Q37	53	65	41	55	46	58	45	52
Q38	36	33	32	39	19	14	28	33

d (JNCS)

Survey Year	1978	1983	1988	1993	1998	2003	2008	2013
#KS	KS6	KS7	KS8	KS9	KS10	KS11	KS12	KS13
Q36	19	24	–	29	30	34	36	45
Q37	53	59	–	65	61	62	62	67
Q38	26	31	–	38	33	33	30	36

e (Seven Country Survey)

Survey Year	1988	1987	1987	1987	1992	1993	1988	1988
	USA	France	UK	West Germany	Italy	The Netherlands	JapanA	JapanB
Q36	54	19	53	43	21	32	31	29
Q37	56	36	58	55	30	48	53	56
Q38	42	23	36	38	14	48	39	34

(continued)

Table 4.5 (continued)

f (Breakdown of Singapore data by Ethnicity)

Survey year	2004			2007			2012		
	Chinese	Malay	Indian	Chinese	Malay	Indian	Chinese	Malay	Indian
Q36	50	54	55	50	57	43	48	58	53
Q37	53	50	53	50	53	50	48	55	48
Q38	33	37	28	25	25	28	33	33	45

g (Japanese Immigrant Surveys)

Survey year	1988		1999		1998	1991
	Hawaii JA	Hawaii Non-J	Hawaii JA	Hawaii Non-J	JAWCS	BRZ JB
Q36	58	66	68	65	66	41
Q 37	68	66	–	–	–	–
Q38	60	46	56	59	61	6

JA: Japanese Americans; JAWCS: JA on the West Coast; Non-J: non-Japanese
BRZ JB: Japanese Brazilian in Brazil

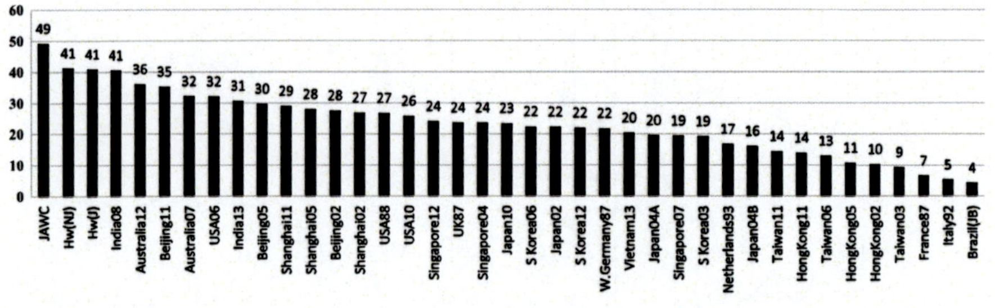

Fig. 4.3 The Percentages (%) of Positive Responses to Both Q36 & Q38 (GSS Trust Items)

people in Latin countries (France and Italy). The percentages of positive response in the United States and UK were high, whereas those in Italy and France were low.

This may look consistent with Fukuyama's [25] theory contrasting Japan, the United States, and Germany as highly trustful countries with China and Italy as less trustful countries. His arguments are based on the assumption that the former countries have well-developed intermediate civic organizations between the government and families whereas the latter have established atmosphere of strict political centralization in the past long histories.

A close look at this figure, however, shows a more complicated reality because the percentages of positive responses of Mainland China (Beijing and Shanghai) were higher than might have been expected by the Fukuyama's argument. There may be several possible explanations for this. First, the data really do indicate that the Chinese have a higher sense of interpersonal trust. (Probably they may be focusing on in-group relationships when responding to those questions). Second, the Chinese respondents might have tried to show a higher sense of interpersonal trust because they were sensitive to their international reputation, such as Fukuyama's argument. Third, the questionnaire items were constructed as a trust scale for Americans, so they may not be suitable for the measurement of trust in other nations. Fourth, we need to be careful about the political and sociological implications of the trust scale. For example, Dogan ([26], p. 258) states, "Erosion of confidence is first of all a sign of political maturity. It is not so much that democracy has deteriorated, but rather the critical spirit of most citizens has improved." This suggests that we need to distinguish between the face value of a scale and its implications. In this context, trust and distrust may not be opposites on a unidimensional scale but instead may be closely related in a sort of multidimensional mind structure. Furthermore, people may give the same response for different reasons or different responses for the same reason. Therefore, for a more meaningful comparison of countries, it is necessary to consider peoples' responses with objective measures on, e.g., economy and politics, as well as general response tendencies of those peoples (Sect. 3.4 of Chap. 3). (Therefore, in this book, I may present some interpretations of the response patterns for certain questions, but they should necessarily be considered tentative).

Lastly, I note on general response tendencies [24]. Yoshino and Osaki [16] and Yoshino [27] reviewed our past surveys on trust and subjective well-beingness, and concluded that the long-term tendency is relatively stable over time regardless of objective economic or political conditions (cf. [28]) although a serious incident or disaster perturbs the stability.

As for general response tendency of people in each country, for example, the Japanese tend to avoid polar answers and prefer intermediate response categories (or "Don't Know"), the French tend to choose critical categories, and the Indians tend to choose optimistic categories (Sect. 3.4 of Chap. 3). Furthermore, as to gender differences, women show stronger self-disclosure than men ([29], Sect. 2; Yoshino, Hayashi and Yamaoka [30], pp. 108–111). This may lead a superficial contradiction, e.g., the women show higher sense of satisfaction when asked about their satisfaction whereas they show higher sense of dissatisfaction when asked about their dissatisfaction.

4.2.3.2 Trust of Social Institutions and Systems

The questionnaires of APVS included the same question items on institutional trust used in the World Values Survey, with an additional item on trust in science and technology. The questions are stated as follows:

Q52. How much confidence do you have in the following? For each one, could you tell me how much confidence you have in them? Is it a great of confidence, quite a lot of confidence, not very much confidence or none at all?

	A Great Deal	Quite A Lot	Not Very Much	None
a. Religious organizations	1	2	3	4
b. The law and the legal system	1	2	3	4
c. The press and television	1	2	3	4
d. The police	1	2	3	4
e. National federal bureaucracy	1	2	3	4
f. National Assembly (Federal Parliament)	1	2	3	4
g. NPO/NGO (non-profit and non-governmental organization)	1	2	3	4
h. Social welfare facilities	1	2	3	4
i. The United Nations	1	2	3	4
j. Science and technology	1	2	3	4

Table 4.6a-c shows the response distribution of the questions. To reduce the effects of general response tendencies particular to countries, Yoshino [31] transformed the response data from the East Asia Values Survey (EAVS) (2002–2005) into standardized scores country by country. Here "general response tendencies" mean, for

Table 4.6 **a.** Insitutional Trust of WVS Items (APVS). The figures show percentages (%) of sum of positive categories "1. a great deal" and "2. quite a lot " . **b**. Insitutional Trust of WVS Items (PRVS). **c**. Insitutional Trust of WVS Items (EAVS).

a (APVS)

	Survey year	2011	2011	2011	2011	2011	2012	2012	2012	2013	2010	2013
	Item	Beijing	Shanghai	Hong Kong	Taiwan	USA	South Korea	Singapore	Australia	India	Japan	Vietnam
Q52a	Religious organization	27	36	58	75	58	41	82	44	87	13	79
Q52b	The law and the legal system	85	82	86	53	56	51	83	78	78	72	94
Q52c	The press and television	68	64	56	44	21	63	78	33	70	70	80
Q52d	The police	75	72	69	59	65	46	92	89	59	70	85
Q52e	National government bureaucracy	83	77	50	46	26	34	89	47	49	38	93
Q52f	Congress/diet	83	75	53	38	22	17	88	46	55	25	95
Q52g	NPO/NGO	41	45	64	56	51	42	80	74	61	49	86
Q52h	Social welfare facilities	78	70	81	69	48	59	83	79	75	71	88
Q52i	The United Nations	59	54	70	61	40	68	82	63	60	59	90
Q52j	Science and technology	95	89	85	86	76	75	91	92	90	83	95

(continued)

Table 4.6 (continued)

b (PRVS)

	Survey year	2005	2005	2005	2006	5006	2006	2007	2007	2008	2004
	Item	Beijing	Shanghai	Hong Kong	Taiwan	USA	South Korea	Singapore	Australia	India	Japan A
Q50a	Religious organization	24	32	60	66	69	46	84	48	90	15
Q50b	The law and the legal system	84	86	87	63	79	57	96	72	90	79
Q50c	The press and television	61	66	46	32	44	59	81	32	82	74
Q50d	The police	72	77	78	57	80	50	95	83	64	69
Q50e	National government bureaucracy	81	85	63	42	45	29	92	46	58	41
Q50f	Congress/diet	82	86	61	25	51	11	90	51	64	32
Q50g	NPO/NGO	40	46	64	40	65	31	70	74	69	45
Q50h	Social welfare facilities	73	74	81	56	62	59	78	74	83	70
Q50i	The United Nations	59	59	65	46	55	58	74	59	75	56
Q50j	Science and technology	96	94	88	73	87	75	93	88	93	76

(continued)

Table 4.6 (continued)

c (EAVS)

2002	2002	2002	2003	2003	2002	2003	2003	2004
Japan	Beijing	Shanghai	Kunming	Hang-zhou	Hong Kong	Taiwan	South Korea	Singapore
12	20	29	24	24	59	64	49	82
74	82	92	78	80	86	65	59	93
74	57	71	55	60	41	37	60	82
64	71	74	64	73	75	50	49	93
41	84	88	71	81	56	42	30	89
34	85	87	72	82	55	30	14	86
55	36	53	39	48	59	38	41	76
68	70	80	61	69	77	52	62	84
62	59	62	45	56	63	50	50	71
73	97	97	95	90	84	76	78	87

example, the Japanese tend to avoid polar responses than the Americans (variances of the range of responses). But let us use an easier way that Yoshino [29] used for the Pacific-Rim Values Survey (PRVS) (2004–2009). First, the original response categories are re-categorized to sum up the percentage of responses to positive categories ("1" and "2"). Second, the percentages of positive responses are compared question by question within each country. This yields a rank order of questions in each country. Third, the rank orders of all countries involved are compared. This procedure results in the loss of some information from the original data, but it may provide more stable cross-national comparability (unless the rank orders are unstable).

Yoshino ([15], Table 3a, b, & c) confirms the generally consistent patterns in the countries or areas participating in all three surveys of APVS, PRVS, and EAVS. For example, the question-by-question differences of percentages between the Japan surveys in 2002 (EAVS), 2004 (PRVS), and 2010 (APVS) were almost within the margin of the sampling error. The maximum difference was about 10%, for example, on non-profit organization/non-governmental organization (NPO/NGO). (The percentage on NPO/NGO changed from 55% in 2002 to 45% in 2004 and then up to 49% in 2010. NPO/NGO activities had been increasing and some disguised NPO/NGOs had managed illegal businesses in the early 2000s. This was one of the reasons that the Japanese laws on registered organizations were substantially revised in 2008).

In Table 4.6a–c, confidence in religious organizations was low in all countries or areas surveyed except for India, Singapore, Vietnam, the United States, and Hong Kong. Even in these five countries or area, the relative degrees of confidence were not very high compared with all the other questions for each country, except for India. Japan and Mainland China indicated remarkably negative attitudes toward religious organizations. The percentage of positive responses among the Japanese was lower than among the Chinese. However, of the 10 questions on Q50, the percentage of positive responses was the lowest for religious organization among Chinese. As already mentioned, most of the Japanese respect religions or the "religious heart/mind" even when they do not have religious faith (Sect. 2.4.8 of Chap. 2 of this book, Hayashi and Nikaido [32]). However, some religious groups such as "Aum Shinrikyo" (a religious cult) caused a serious disaster in the 1990s, so the Japanese may keep cautious to "religious groups." In China, the government is very sensitive toward religious groups because, in the long history of China, religious groups frequently overthrew governments. In some countries, some religious groups are closely linked to terrorism.

The percentages of responses that show confidence in authority such as the "police," "government," and "Congress" may represent various patterns of attitudes. These are likely concerned with democracy. Because free criticism is allowed in democracy, a negative attitude does not necessarily mean the negation of such authority, and it may reflect a mature democracy in some countries ([26], p. 258). Thus, the percentage of positive (or negative) responses may not be linearly proportional to the degree of political maturity. Table 4.6a-c shows, for example, USA's lower degrees of confidence in the press and TV and in Congress. This may be a critical attitude of matured democracy, or it may be a reflection of current confusions of democracy, or both.

Hayashi [1] and Zheng and Yoshino [33] presented cross-national analyses of data on science and technology from our Seven-Country Survey. As explained in Sect. 4.1.4 of Chap. 4, Hayashi [1] concluded that the Japanese generally have positive attitudes toward science. They were, however, negative regarding scientific approaches toward the understanding of the human heart and mind ("kokoro" in Japanese), solving social and economic problems, and the possibility of living in space stations in the near future (at the time of the survey in 1988). The response pattern of West Germans in 1987 was similar to that of Japanese in the sense that they were also more negative about science and technology than those in other Western countries. However, they were not so negative toward the applications of science and technology to social problems as well as psychological problems of individuals as the Japanese were. This may be related to the fact that Hegel or Marx's social theory and Freud's psychological theory began in the German cultural sphere.

As for data from APVS, all of the countries or areas were highly positive toward science and technology, with rates of positivity for that question being the highest among all questions. In particular, the rates for Mainland China were remarkably close to 100% in both PRVS and EAVS, although the rates were slightly down in APVS. There may be several possible explanations for this. On the one hand, the high rates may represent the fact that, since the late 1970s, China has been emphasizing the scientific reformation of government agencies, military systems, and social systems as a priority in their social planning. On the other hand, until recently, they had placed priority on economic development and they had not paid much attention to the negative impact of science and technology that advanced industrial countries have experienced in the past. After the Beijing Olympics in 2008 or even slightly earlier, the Chinese government started paying attention to the negative side of rapid economic and industrial development and began planning to improve environmental conditions, including serious air, soil, and water pollution. Incidentally, they started also paying attention to political issues, such as the social inequality between urban and rural areas. They are struggling to deal with these domestic problems, but complete solutions seem still far away to many observers' eyes [34, 35], despite their rising power in international politics. Furthermore, under the recent slowdown of economic development, environmental improvement may not be on the government top priority. And serious pollutions in the urban areas sometimes go over to the neighboring areas and countries. More recently, however, certain steady environment improvements in some rural areas are reported [36].

As a final comment in this section, it should be noted that Sasaki and Suzuki ([37], Chap. 11) concluded that "a single scale is not adequate to measure people's sense of trust in science and technology because people's attitudes differ from one issue to another within the fields of science and technology." This is also the case with our study on people's sense of trust in general. Note that we have Japanese Nobel Prize Laureates in the 1990s–2010s more than in the past. The data of JNCS [2], pp. 156–157 [questions #9.6 & #9.12]), however, shows a clear decrease of self-confidence in science and technology during "the lost decade (1993–2003 or so)." Thus, while generalized interpersonal trust is more stable, some aspects of confidence are more volatile due to economic or political conditions.

4.2.4 Sense of Trust Among Japanese Immigrants and Ethnic Differences

Domestic ethnic differences on trust are often linked with domestic and international political issues. Mutual trust is a key for solving those issues.

In this section, I summarize regional and generational differences of Japanese immigrants in Hawaii [9], Brazil [38], and the U.S. West Coast [12], as well as ethnic differences (the Chinese, Malays, & Indians) in Singapore and between Taiwanese and Chinese mainlanders in Taiwan.

Our past surveys have already clarified similarities and dissimilarities between Japanese living in Japan and Japanese immigrants abroad. First, regarding the Japanese tendency to avoid polar categories and to prefer intermediate response categories in a questionnaire survey, there are some variations even between Japanese-Americans in Hawaii and on the West Coast in the United States [3], Hayashi et al. [6], p. 388; Sect. 3.4 of Chap. 3 of this book). This may be a result of the situational differences between the place where they need to make their intentions clear to communicate successfully in their host countries (e.g., the mainland USA) and the place where the Japanese comprise the largest group among the minorities (e.g., Hawaii) (see Note 4).

Second, we found that the Japanese style of interpersonal attitudes ("giri-ninjyo" attitudes) or religious attitudes (the Japanese consider religious mind/spirit to be important, whether or not they have religious beliefs) was preserved among the immigrant first and second generations in the United States. The third and younger generations seem to have become more adapted to the country where they were born and now live (Hayashi [1]; Yamaoka [39] Fig. 4, Table 3; ; Yoshino [22]).

Third, Brazilian-Japanese ([38], p. 435) showed a much lower degree of sense of trust for each of the three questions (Q36, Q37, and Q38) than most responders in the other countries (Table 4.5g) (see Note 5). Furthermore, an even lower degree of trust was found in the third or younger generation than in the first and second generations . As a whole, the pattern of sense of trust in Brazil appeared close to those of France (1987 survey) and Italy (1992 survey). As for the immigrants in Hawaii and on the West Coast (Table 4.5g), they showed a higher degree of trust than most responders in the other countries, including Japan and the United States. (Unfortunately, the survey of the US West Coast Japanese immigrants did not include Q37).

For an understanding of the generations of Japanese immigrants in the United States, there is a frame of comparison, as Kitano ([40], Chap. 13) showed. The first generation was called "Japs" and was discriminated against as immigrants from Japan. The second generation represents Japanese-Americans who have been constantly in search of their own identity between the United States and Japan (many of them voluntarily fought in World War II as American soldiers, while their parents were placed in concentration camps by the American government). The third generation or younger are simply Americans of Japanese descent.

Among Honolulu residents, there was no significant difference in patterns of interpersonal trust between Japanese-Americans and non-Japanese Americans (Q36, Q37, and Q38) ([22], Fig. 8). (In Hawaii, no ethnic group was the majority, and Japanese-Americans comprise the largest group among the minorities). In addition, Yoshino ([22], Fig. 9) provided details of ethnic comparisons between Hawaiians, but each sample size is not sufficient for statistically reliable comparisons.

Time series data from the Hawaii Resident Survey showed longitudinal changes among Japanese residents of Hawaii in terms of trust for each generation ([22], Fig. 10). There was no large difference between generations over the two decades, except for some differences between second and third generations in 1988. This might be explained by the complicated mixture of ethnicities in Hawaii. In addition, there was little change in the sense of interpersonal trust in each generation over those two decades (1978–1999).

Data for the West Coast showed a large difference between generations: the older generation seemed to be more trustful ([22], Fig. 11), although this was not seen in the Hawaii data. We should be very careful when interpreting the data because there may be several compounding factors (such as age, generation, physical condition, economic conditions, and residential area). The difference between Japanese Americans in Hawaii and Japanese Americans on the West Coast with regard to generational differences may be related to differences in their concepts of ethnicity. That is, Japanese-Americans on the mainland would have no problem classifying themselves as Americans of Japanese descent. Conversely, many Hawaiian residents are of highly mixed ethnicity in this multiethnic society, so people's perception of ethnicity depends upon how they identify themselves in the ethnic classification—this ought to be no different for the Japanese-Hawaiians. The different conditions of Japanese-Americans in Hawaii and on the West Coast during World War II may have had an effect on their attitudes and social values (Note 4).

Next, see Table 4.5f on the ethnic differences (Chinese, Malays, and Indians) in Singapore. General patterns seem stable over time, although there may be some changes in the response distributions for the questions. Among others, an increase in the positive response to Q38 among the Indians (generalized interpersonal trust) may be remarkable: 28% in both 2004 and 2007 and 45% in 2012. This increase might be related to some changes in Indian immigrants' status in Singapore caused by India's rapid development in the international economy and politics in the last two decades. We might need to be careful, however, about the possibility of statistical fluctuation because of the smaller sample size that accompanies an analysis involving breaking down the sample by ethnicity. This is also the case with the ethnic differences in Singapore shown in Table 4.5f.

4.2.5 Universal Social Values on Human Bonds

People's sense of trust does not come in a vacuum. To overcome the limitations of studies mostly based on response distributions of the questions of GSS or WVS, it may be important to explore more basic social values that may underlie people's sense of trust. Although it is not easy to identify basic social values in various countries, here I tentatively show some data on Confucian teachings, several important areas of daily life, and choices of Asian values versus Western values as follows.

Very strictly, there is a distinction between Confucian teachings and Confucianism. The former was preached by Confucius around the sixth century B.C. for people's daily life, whereas the latter has been organized by scholars much later after his death for maintenance of order of centralization of governmental politics (Seki [41]). Confucius teachings and Confucianism are quite different from Christianity and Islam, but at least in some places and times they were effective as a kind of religion or ethics to maintain people's daily lives and social order. Therefore, it would be worthwhile to investigate it for our study of trust. Here, not necessarily sticking to the above-mentioned historical distinction, I deal with some questions for people's daily life regarded as basic Confucius teachings. The questions are originally from a comparative survey of Tokyo (Japan), Seoul (South Korea), and Beijing (China) as an international collaboration of Mainichi Newspaper July 1997.

First, Table 4.7a, b, c, and d show the response distributions with regard to Confucius teachings from APVS and the related surveys. The question asked is as follows (see Note 6).

> Q9. How do you feel about each of the following traditional values? (The response categories are "strongly agree," "agree to some extent," "disagree to some extent," and "disagree strongly.")

a. We should respect our ancestors.
b. The eldest son should look after his aging parents.
c. A wife should obey (follow) her husband.
d. One should not marry someone his/her parents object to.
e. We should obey (follow) older people.
f. It is important to have a son to keep the family line going.
g. Men should work outside the home and women should tend to housekeeping.

The patterns of response distributions have been consistent for the countries or areas surveyed in EAVS [31], PRVS [29], and APVS [15], although the percentages may show some changes over time. For some questions, the percentages of positive choices of the United States are larger than those of the Asian countries. It cannot be the case that Americans are more influenced by Confucian teachings than Asians. Overall, Asian countries already seem to be more or less distant from the literal teachings of Confucius.

More detailed analyses, however, show that there are national differences between, e.g., China, South Korea, and Japan, as well as gender and age differences (cf. Zheng

Table 4.7 **a.** Confucian Teachings of the Asia-Pacific Values Survey (2010–14). The figures show percentages of sum of positive categories "1.strongly agree" & "2.agree to some extent". (Note 1: In USA 2006 & Singapore 2007 surveys, the word "follow" was used in Q9c & Q9e by mistranslation.) (Note 2: This table gives a comparison of chinese mainlander and nativeTaiwanese in 2006, but it is becoming diffficult to ask about this distinction these days.) **b.** Confucian Teachings of the Pacific—Rim Values Survey (2004–08). **c.** Confucian Teachings of the East Asia Values Survey (2002–05). **d.** Breakdown by Social Group or Ethnic group

a (APVS)

Item	2011	2011	2011	2011	2012	2012	2010	2012	2012	2013	2013
	Beijing	Shanghai	Hong Kong	Taiwan	South Korea	Singapore	Japan	Australia	USA	India	Vietnam
Q9a We should respect our ancestors	99	97	95	99	82	96	92	95	97	95	100
Q9b The eldest son should look after his aging parents	77	74	79	68	27	55	37	39	54	86	77
Q9c A wife should obey her husband	33	27	44	43	49	66	34	26	41	86	77
Q9d Not to marry someone whom your parents object to	11	16	10	14	49	40	27	19	34	86	43
Q9e We should obey older people	61	58	61	61	60	82	56	54	62	96	94
Q9f It is important to have a son to keep the family line going	31	27	31	38	36	47	33	29	51	74	73
Q9g Men should work outside and woman should tend to housekeeping	30	28	30	31	32	29	24	13	22	66	44

(continued)

Table 4.7 (continued)

b (PRVS)

Survey Year	2005	2005	2005	2006	2006	2007	2004	2007	2006	2008
Item	Beijing	Shanghai	Hong Kong	Taiwan	South Korea	Singapore	Japan	Australia	USA	India
Q9a We should respect our ancestors	96	97	99	98	84	96	91	96	96	100
Q9b The eldest son should look after his aging parents	70	61	76	66	45	55	32	44	45	77
Q9c A wife should obey(follow) her husband	21	20	41	45	60	60	36	29	52	82
Q9d Not to marry someone whom your parents object to	7	12	7	11	44	28	27	18	37	76
Q9e We should obey (follow) older people	43	50	54	66	72	51	54	44	56	96
Q9f It is important to have a son to keep the family line going	21	25	34	39	34	43	30	35	45	74
Q9g Men should work outside and woman should tend to housekeeping	21	20	29	40	26	28	26	15	17	56

c (EAVS)

Survey Year	2002	2002	2003	2002	2002	2003	2003	2004	2002
Item	Beijing	Shanghai	Kunming	Hangzhou	Hong Kong	Taiwan	South Korea	Singapore	Japan
Q9a We should respect our ancestors	96	97	93	94	97	100	88	96	92
Q9b The eldest son should look after his aging parents	50	51	51	59	64	67	45	41	31
Q9c A wife should obey her husband	12	11	15	14	40	55	64	65	32
Q9d Not to marry someone whom your parents object to	3	6	5	6	7	18	32	25	22
Q9e We should obey older people	37	40	30	32	49	68	62	46	43
Q9f It is important to have a son to keep the family line going	15	11	15	17	30	53	41	42	28
Q9g Men should work outside and woman should tend to housekeeping	12	14	16	12	25	46	35	29	23

(continued)

Table 4.7 (continued)

d (Breakdown by Social Group or Ethnic Group)

Survey Year	Taiwan	2006	Singapore	2004		2007				2012			
Item	Native Taiwanese	Chinese mainlander	Malay	Indian	Others	Chinese	Malay	Indian	Others	Chinese	Malay	Indian	Others
Q9a We should respect our ancestors	98	96	94	98	96	96	94	96	94	96	96	94	–
Q9b The eldest son should look after his aging parents	65	70	54	51	44	51	66	80	63	43	65	59	–
Q9c A wife should obey(follow) her husband	44	50	83	72	63	56	81	68	66	61	84	74	–
Q9d Not to marry someone whom your parents object to	12	9	44	49	37	24	40	40	28	35	55	50	–
Q9e We should obey (follow) older people	67	48	65	70	70	45	69	69	63	79	88	92	–
Q9f It is important to have a son to keep the family line going	40	35	55	46	48	41	51	41	47	47	48	50	–
Q9g Men should work outside and woman should tend to housekeeping	40	37	39	24	33	27	31	26	28	28	37	32	–

[42]). Confucius teachings or Confucianism originated in China, but since the end of fourteenth century A.D. it has a greater influence in Korea than in China. On the other hand, it had an influence limited to the dominant class in Japan only during the Edo era ("samurai era" of the seventeenth to the nineteenth centuries). These historical differences may have led to national differences in the response distributions for those questions. It is important to note that, under the present conditions, people cannot follow Confucian teachings literally. For example, the decreasing number of children in almost all countries in East Asia is a serious problem for the maintenance of national productivity or social systems concerning health care and financial support for older people. Thus, it is becoming difficult for elderly parents to be completely taken care of by their children, not to mention by the eldest son alone (Q9 b).

Table 4.7d also shows a comparison of native Taiwanese ("Honsyo-jin," i.e., the Taiwanese who have been in Taiwan before World War II and their offspring) and the mainlanders in Taiwan ("Gaisyou-jin," i.e., the Chinese who have come from the mainland after World War II and their offspring). Almost the same patterns are seen, except on Q9e (obedience to seniority): more positive responses are seen among the native Taiwanese than the Chinese mainlanders. However, we must be careful regarding the distinction of native Taiwanese and Chinese mainlanders. Although the categories are important for analyses, they occasionally relate to sensitive political matters concerning the relationship between Taiwan and the mainland of China. In the survey, the distinction between the two groups is based on self-reporting. And, as in the usual case of rigorous statistical random sampling surveys, the survey cooperation rate did not reach 100%. As a result, in our Taiwan survey, there were fewer "Gaisyou-jin" than we would expect from census data. There are at least two possibilities: our sample might have been truly biased, or the mainlanders might have hidden their identities for political reasons. Therefore, the aforementioned interpretation should be considered tentative. However, this does not necessarily mean that the quality of the data is poor. Rather, it is important to observe how people react to our type of survey, before jumping to the numerical analysis of responses in questionnaire. For some politically sensitive countries, social desirability may be remarkable.

Here, it may be worthwhile to note that the unexpected combination of Confucian teachings and advanced medical technology caused an ethics crisis in South Korea and Taiwan around the 1990s or so. As seen in questions "b" and "f" of Q9, there is a strong social pressure on women to bear sons under Confucian teachings. Wives had a hard time when they did not bear a son, and in some cases, they adopted a boy from their relatives. However, modern medicine has made it possible to bear a child by a surrogate mother. In several cases, however, rich South Korean parents did not honor their responsibility to take care of the babies when they found that surrogate mothers (occasionally poor Chinese women) were to give birth to female babies. Another issue with fertility rates is that the proportion of men and women in China and South Korea is significantly disproportionate (more men than women). This may be pieces of evidence against the "utility" of modern medicine without new ethics that are suitable for modern science.

Incidentally, the adoption of children is different among Japan, China, and Korea: the Japanese may adopt a child even if there is no blood relationship between the child and the parents, whereas the Chinese and Koreans adopt a child from among their relatives whenever possible.

Section 3.4.2 of Chap. 3 presented Q19 on of the importance of various areas in daily life. Table 4.8a, b, c, and d shows for each question and each country or area the sum of the percentages of positive categories 5, 6, and 7 on the response distributions. In all the countries we have surveyed, the degree of importance of "immediate family" was the highest among the various aspects of daily life. The degree of importance of "relatives" was the second highest, but it may show some variations compared to those of "immediate family": the percentages of positive categories for relatives were closer to those of "immediate family" for Asian people as well as Japanese immigrants than for Western people (except Italians). These findings may be closely linked to national character. India, Singapore and Vietnam show high degree of confidence in religious organizations in the countries of APVS (Table 4.8a), but all the three countries show a higher degree of importance for "family" and "relatives" than for "religions." Thus, we confirm that all people think that family and relatives are important among those aspects of daily life, disregarding differences between East and West, cultures, or religions.

As for the third highest choice, the percentages selecting "career," "friends," or "free time" varied among countries. This seems to be related more to economic conditions than to national character, so the percentages may vary more over time.

Next, the Seven-Country Survey, EAVS, PRVS, and APVS included the following question, which asks respondents to choose two response categories from the four presented: two Asian values (originally from Confucian teachings) and two Western values (originally from "United States Declaration of Independence" or "Déclaration des droits de l'homme et du citoyen").

Q34 [SHOW CARD] If you were asked to choose the two most important items listed on the card,

which two would you choose? (Select two).

a. Filial piety/love and respect for parents
b. Repaying people who have helped you in the past
c. Respect for the rights of the individual
d. Respect for the freedom of the individual

Six possible pairs can be selected among the four items. Table 4.9a, b, and c shows the percentages of the selection of each possible pair in the surveys. Some countries, such as Japan and the United States, were studied in all the three surveys. The patterns selected in Japan are fairly consistent, whereas those of the United States differ slightly. (The 1988 Japan survey and the 2004 Japan survey used the same fieldwork company and sampling method. The fieldwork firm of the 1988 US survey was Gallup, and that of the 2006 US survey and the 2010 US survey was Kane, Parsons & Associates. Gallup used random-walk sampling, and Kane, Parsons & Associates used quota sampling with respect to age, gender, and race in 2006 and random-route sampling with a quota table on age, gender, and race in 2010.

Table 4.8 **a**. The Degrees of Importance of Aspects in Daily Life: The Asia-Pacific Values Survey (APVS). The figures show the sum of percentages (%) of positive categories 5, 6 & 7. **b**. The Degrees of Importance of Aspects in Daily Life: The Pacific-Rim Values Survey (PRVS). **c.** The Degrees of Importance of Aspects in Daily Life: The East Asia Values Survey (EAVS). **d.** The Degrees of Importance of Aspects in Daily Life: The Sevenn Country Survey

a (APVS)

survey year	2010	2011	2011	2011	2011	2012	2012	2013	2012	2010	2013
	Japan	Beijing	Shanghai	Hong Kong	Taiwan	South Korea	Singapore	India	Australia	USA	Vietnam
Q18a Family	97	98	96	96	98	99	95	97	97	94	98
Q18b Career	89	90	92	90	93	95	93	96	83	86	99
Q18c Free time	92	81	86	92	95	98	98	78	93	90	88
Q18d Friends	96	91	93	94	95	99	95	92	94	95	84
Q18e Relatives	96	97	96	95	98	98	99	96	95	95	97
Q18f Religion	42	26	42	50	64	65	80	81	48	77	58
Q18g Politics	78	60	57	52	38	69	43	38	41	51	60

b (PRVS)

survey year	2004	2005	2005	2005	2006	2006	2007	2008	2007	2006	1988	
	Japan A	Beijing	Shanghai	Hong Kong	Taiwan	South Korea	Singapore	India	Australia	USA	Japanese Americans	Non-Japanese
Q18a Family	93	93	92	92	97	97	96	96	92	91	95	94
Q18b Career	72	77	85	80	89	96	86	93	66	77	73	80
Q18c Free time	70	59	69	73	85	94	87	68	80	82	78	75
Q18d Friends	81	67	80	74	80	91	86	82	83	88	87	80
Q18e Relatives	89	89	89	90	95	96	97	93	89	93	94	85
Q18f Religion	21	14	28	31	38	47	69	81	36	68	61	58
Q18g Politics	49	43	51	23	19	43	39	37	29	35	39	39

(continued)

Table 4.8 (continued)

c (EAVS,and Japanese Resident Surveys in Hawaii, the West Coast of USA and Brazil))

survey year	2002	2002	2002	2003	2002	2002	2003	2003	2004	1999	Hawaii 1999-	JA on the W.Coast 1998	Japanese Brazilians 1991
	Japan	Beijing	Shanghai	Kunming	Hangzhou	Hong Kong	Taiwan	South Korea	Singapore	Japanese Americans	Non-Japanese	1998	1991
Q18a Family	94	95	98	93	95	94	97	97	95	99	94	JAWCS	BRZ JB
Q18b Career	78	87	91	84	90	81	88	95	81	66	77	93	94
Q18c Free time	77	68	74	70	78	74	84	93	88	80	80	74	95
Q18d Friends	85	76	82	77	84	76	76	93	88	91	83	83	63
Q18e Relatives	92	91	94	91	94	92	94	97	97	97	88	92	87
Q18f Religion	24	11	18	17	24	28	43	49	72	58	55	96	96
Q18g Politics	48	41	44	46	45	20	22	45	37	34	31	47	74

d (Seven Country Survey)

	Japan-A	Japan-B	FRA	UK	FRG	ITA	HOL	USA
Q18a Family	95	94	93	96	87	98	94	98
Q18b Career	85	83	85	57	51	75	62	65
Q18c Free time	76	78	73	66	83	70	83	74
Q18d Friends	87	85	73	76	84	80	88	83
Q18e Relatives	91	91	83	83	81	93	86	90
Q18f Religion	38	37	35	36	38	64	33	76
Q18g Politics	54	53	23	26	40	28	32	46

Table 4.9 a. The Rates (%) of Two Choices from the Four Choices (Q34). APVS A (filial piety), B(pay back to a benefactor), C(the right of individuals) & D (the right of freedom). b. The Rates (%) of Two Choices from the Four Choices (Q34). PRVS A (filial piety), B(pay back to a benefactor), C(the right of individuals) & D (the right of freedom). c. The Rates(%) of Two Choices from the Four Choices (Q34). The Seven Cvountry Survey. A (filial piety), B(pay back to a benefactor), C(the right of individuals) & D (the right of freedom)

a (APVS)

Survey year	2010	2010	2011	2011	2011	2011	2012	2012	2013	2012	2013
	USA	Japan	Beijing	Shanghai	Hong Kong	Taiwan	South Korea	Singapore	India	Australia	Vietnam
A&B	**21**	**55**	**67**	**57**	**44**	**54**	**43**	**48**	**39**	18	**37**
A&C	**22**	10	15	18	15	17	14	27	29	**26**	27
A&D	14	13	12	12	22	13	17	10	22	12	27
B&C	15	6	2	3	5	6	7	6	4	11	4
B&D	6	11	2	3	6	5	8	3	4	5	2
C&D	19	5	2	3	7	5	10	7	2	**27**	4

b (PRVS)

Survey year	2006	2004	2005	2005	2005	2006	2006	2007	2008	2007
	USA	Japan	Beijing	Shanghai	Hong Kong	Taiwan	South Korea	Singapore	India	Australia
A&B	**26**	**48**	**55**	**58**	**44**	**53**	**51**	**45**	**34**	16
A&C	23	12	19	18	14	17	18	30	23	**31**
A&D	11	15	13	12	26	12	15	12	11	12
B&C	12	8	5	4	3	5	7	6	2	8
B&D	7	12	4	3	7	7	6	3	2	4
C&D	22	8	5	4	7	6	4	4	1	25

(continued)

Table 4.9 (continued)

c (Seven Country Survey)

Survey year	1988	1988	1992	1987	1987	1993	1987
	USA	Japan	Italy	France	W.Germany	Holland	UK
A&B	20	**47**	24	20	9	12	**32**
A&C	**40**	14	**30**	14	28	**30**	20
A&D	13	20	26	20	19	26	13
B&C	6	5	4	8	4	3	11
B&D	3	7	3	11	3	1	8
C&D	20	8	14	**27**	**37**	**28**	16

Note Table 6b of Yoshino ([15]) had typographical errors in the survey years, but it was corrected in the above Table 4.9b.

Therefore, it is not immediately clear if the slight change in the response distribution in the United States is real or due to differences in sampling companies or methods).

In all countries or areas in Asia, the first choice was the pair "a. Filial piety/love and respect for parents" and "b. Repaying people who have helped you in the past." In contrast, even in Western countries, the choice of the pair "c. the rights of individuals" and "d. the right of freedom" was not a majority choice. Only in France and West Germany (in 1987 before the reunification) was the rate of selection of that pair higher than those of the other pairs, but at most, it was 30% or so.

For almost all countries, "a. Filial piety/love and respect for parents" was the single first choice, i.e., the rate of the selection for each of the possible four choices tallied individually, even though Q34 asked the respondents to choose two items, not just one (see Note7); this is consistent with the previous observation from data on the importance of daily life Table 4.8 a-d. That is, the degrees of importance of immediate family and relatives were the highest and the second highest, respectively, in all countries or areas. Japanese generally tend to prefer the medium answer category, but for issues of family importance, they also choose the highest scale points. Incidentally, it may be interesting to note a theory of Todd ([43]) on family structures and political regimes, which suggests that the political regime is a reflection of family structure in each country.

No one today denies the importance of individual rights and the right to freedom. However, they have only recently appeared in history, and throughout the long history of mankind, "family and relatives" have laid the foundation for people's survival. Family and relatives are probably universally valued in both the East and the West, whereas the rights of individuals and the right to freedom originated in the "modern West," so they may be viewed differently in the current political system of each country. (Universal social values can also be seen in other aspects such as "work." However, it can also be related to interpersonal and social relationships through work. See Note 8).

4.2.6 For Future Study of Trust

Regarding senses of interpersonal and institutional trust, I have repeatedly emphasized that we need to be cautious in interpreting the data because survey data on "trust" are often a compound of many variables, including economic and political factors as well as people's general response tendencies. In addition, we must consider the utility and limitation of items that are used for measuring trust. Scales may be used to measure some important dimensions related to trust, but there may be other important dimensions of trust. I will provide a few comments for our future research as follows.

First, in order to facilitate a mutual understanding between the East and the West, we need to pay much attention to measurements of social values. Studies on the scale of trust may caution us on the applicability of a certain "single" scale invented in Western cultures for Eastern cultures, or vice versa. Gallup ([44], p. 461) reported that, in their global survey, they could not find very poor but still happy people. Later studies, however, have found examples not consistent with the pattern of Gallup report. For example, Brazilians were very optimistic even when Brazil was the worst debtor nation in the 1980s [45]. Inglehart reported a correlation of 0.57 between economic development and life satisfaction for some 20 countries surveyed in the 1980s ([45], pp. 366–371). However, life satisfaction in Japan in the 1980s was lower than it was around 2000 or after 2013, although Japan was prosperous in the 1980s but struggled with a recession around 2000. Even more noteworthy is the rise in life satisfaction in a 2011 poll following the Great East Japan Earthquake. The Japanese witnessed many lost lives and the sadness of their bereaved families, recognized the importance of lives, and deeply appreciated their daily lives. In measurement of social values, it is necessary to be careful about peoples' situations and personalities.

Second, it is worth considering Dogan's statement ([26], p. 258). "Erosion of confidence is first of all a sign of political maturity. ... It is not so much that democracy has deteriorated, but rather the critical spirit of most citizens has improved." Some people may express distrust or complaint toward the government or political leaders not because they lack interpersonal trust but because they know that it is a way to improve their own country and eventually our world in a democratic way.

Third, a comment is about Weber (1904–05)'s argument on religion and capitalism that Asian countries would not be able to develop capitalism. Obviously, many counter-examples (such as Japan, South Korea, NIES, and China) have gone against his argument. Some have argued that the Japanese adaptation of Confucianism functioned as a substitute for the Protestant ethic and led Japan to successfully develop capitalism [46]. However, the last century has seen many examples indicating that economic success is not necessarily linked to a particular ethic, ideology, or religion. Now we have more data to consider the relationships among economic development, social systems, and social values.

In this time of globalization, there are various ways to achieve successful social development. We should not impose our own social values on any other country if we intend to develop a peaceful world.

Notes

Note 1. Codes such as Q36 correspond to the common question code of APVS questionnaire. For the precise wordings and response data, see the series of ISM Survey Research Reports (https://www.ism.ac.jp/editsec/kenripo/contents_e.html) or the website for ISM surveys (https://www.ism.ac.jp/ism_info_e/kokuminsei_e.html).

Note 2. As for Q38, there are slight differences in wording between our cross-national Japan survey and the Japanese National Character Survey. In the process of a translation and back-translation check to make a cross-national version of the Japan survey questionnaire, we obtained the two versions. We have confirmed that this difference may produce some percentage differences in the response distributions, but the overall pattern yielded by multidimensional analysis is stable (Sect. 3.2.3 of Chap. 3).

Note 3. This generation more or less overlaps those of "yutori-kyoiku." Usually the "yutori kyoiku" (relaxed education) is referred to primary educational reformation started in the late 1980s, concerning a teaching guideline (Curriculum Guideline). Since the 1970s, the Japanese government gradually reduced the amount of class time and the contents indicated in the guideline. And major revisions of the guidelines were implemented in 1989 and 1998. Around 1999, however, a decline in the academic abilities of university students became a serious concern. Elementary and secondary education started to be reconsidered. This trend focused criticism on the teaching guidelines. Regarding the graduate level of the university, it seems that many professors have noticed a decline in student competence and knowledge level, even in the early 1990s, as the Ministry of Education obliged the university to accept more graduate students.

The "yutori-kyoiku" may have appeared as an ideal policy for the twenty-first century, but it lead to confusion in practice.

Note 4. In general, second-generation Japanese American soldiers experienced emotional conflict with their parents when they went to war as American soldiers. However, Hawaiian Nikkei soldiers are said to have had few such conflicts. In Hawaii, the Japanese were the largest group of ethnic minorities. And the war between Japan and the United States began with the Japanese attack on Pearl Harbor in Hawaii. These may have influenced the difference between the Japanese-Americans living in Hawaii and in the mainland US. For more details, see Dowds [47], Kashima [48], Fugita and Fernandez [49], and Miyamoto [9].

Note 5. The economic condition of Brazil was bad in the year of the survey (1991).

Note 6. A mistranslation was introduced by the survey agent in the United States: "obey" should have been used in "c "and "e" where "follow" was used. However, this mistake itself may show a lack of understanding concerning Confucian teachings in the United States. The mistranslation was left in USA 2006 survey and in the Singapore 2007 survey but was corrected in the surveys of Australia 2007 and India 2008, and the corrected version was used in APVS (2010–2013), including USA 2010 survey and 2012 Singapore survey.

Note7. Exceptions are the surveys in W. Germany 1987, Hawaii and non-Japanese 1988 and 1999, where "c. the right of individuals" and "a. filial piety" showed the highest and the second highest rates, respectively, although the differences are not large (cf. The Ten Commandments of Moses include "Honour thy father and thy mother [Respect your parents])."

The response patterns for EAVS, PRVS, and APVS are almost the same for the countries or areas surveyed in all of these surveys. (See Table 5.1 of Yoshino [31] for the results of EAVS).

One needs to be careful of the wording of category "a. filial piety." The category is originally from Confucius teachings, and the English translation may not completely express its meaning and nuance. The original word implies not only love and respect to one's parents but a sort of behavioral obligation. In the translation process of the Seven-Country Survey, the international version of 親孝行 (filial piety in Japanese) became “Filial piety/Love and respect for parents.” However, in the translation and back-translation process of the East Asia Value Survey, the international version was accurately restored to its original word 親孝行 (filial piety) in some of the Asian countries.

Note 8. Yoshino, Shibai & Nikaido (eds). [50] show all the data from major surveys on the following question: “Q.15 If you were to get enough money to live as comfortably as you would like for the rest of your life, would you continue to work or would you stop working?”.

In those surveys, the majority of every country and every time chose “continue to work,” with the exception of West Germany in 1987 survey. The 1980s were a period of introspection on the balance between leisure and diligence after West Germany showed rapid economic development after World War II.

On the one hand, this question is clearly related to people’s attitudes toward paid work and leisure. But on the other hand, given that people live in social relationships through work, it seems to be related to the deeper social value of work (rather than just labor for wages) across nations and time. (For the data, see p. 34, p. 108, & pp. 126–127 of https://www.ism.ac.jp/editsec/kenripo/pdf/kenripo117.pdf).

4.3 Summary

Our early comparative analysis of Japanese, overseas Nikkei, and Westerners [1] depicted more clearly about many Japanese features. This includes “giri-ninjyo” (a certain emotional conflict of obligation and warmheartedness particular to the Japanese), other social values and attitudes, tradition versus modernization, leadership, science and heart/mind, etc.

On the scales of “giri-ninjyo” and “ninjyo,” Nikkei (Japanese immigrants) are generally closer to Japanese in terms of relationships. However, Japanese-Americans in Hawaii are closer to Americans in that they prefer rule-following bosses to the Japanese favorite “Ninjyo Katyo (humanistic bosses).” But overall, Hawaiian-Japanese and Brazilian-Japanese are generally similar, although Japanese-Brazilians are more like Japanese. These observations may be related to the differences between those Japanese communities in terms of their social distance from other ethnic communities. (A later analysis by Yoshino [22], including data from Japanese-Americans on the west coast of the United States, suggests that the comparison of Japanese immigrants was more complex and their similarities or differences differed from issue to issue).

The CLA paradigm has been applied to the analysis of cultural connections between Japanese, Japanese-Americans, and Americans. In our joint research, Japanese-American researchers Miyamoto, Fujita, and Kashima developed a theory of interpersonal relationships derived from G.H. Mead’s approach. This theory focuses on the balance of people between direct action and deep cognition, with the aim of facilitating an understanding of the differences in personality between

Japanese and Americans. It indicated that the Japanese prefer delayed action and pay more attention to the perceptions of others, of themselves, and of the situation, whereas Americans favor direct action and pay more attention to their own views. In the conjoint research, it was recognized that Japanese-Americans have a dual thinking framework (mindset) as "Americans" and "Japanese-Americans" in their daily lives.

The Japanese way of thinking lasted from the beginning of the Meiji era in 1868 to the late 1970s by learning advanced Western techniques while maintaining the spirit of Japanese tradition. Among the countries compared (Hawaii residents, UK, Germany, France, and Japanese-Brazilian) at least until the 1990s, the idea of contrasting tradition with modernization was unique to the Japanese living in Japan. (However, the mid-1970s saw a sign of collapse of the trend due to a significant generational change).

Besides, the Japanese are generally positive about the usefulness of science, but in contrast to Westerners, they are negative about solving social problems and completely elucidating the human mind through a scientific approach. Germans (West Germany in 1988) are relatively negative about science, but not as negative about social issues and the usefulness of scientific approaches to the human mind.

Regarding preference of leader, the Japanese prefer a judgmental person who is respected and liked by his subordinates (i.e., a person of harmony, moderation, tolerance, and courtesy), while the Chinese prefer a technically superior and determined person who benefits his subordinates (i.e., parental mercy and child filial piety, justice and morality).

Under the rapid economic and political reforms of the "globalization" era of the 1990s, the Japanese witnessed the collapse of basic interpersonal relationships at home, at school, and at work. This may have destroyed Japan's basic social values and attitudes, which were otherwise stable. This seems to be one of the main reasons why Japan was unable to recover in the long term after the "burst of the bubble economy" from 1991 to 1993.

Furthermore, the studies of interpersonal trust, institutional trust, and other social values depicted features of people of several countries and Japanese immigrants overseas. Some universal social values on human bonds such as the importance of family was recognized, although the styles of family may be different across countries or time. On the other hand, these studies have shown that trust scalability, and perhaps general subjective scalability, is not simple across cultures, and that a scale and an object to be scaled are complementary or interactive. As Decline in trust in power can sometimes be a sign of political maturity that the critical spirit of most citizens has improved [26]. Trust and distrust may not be the opposite on a one-dimensional scale. They are closely related to a kind of multidimensional mind structure in each culture and each social condition.

References

1. Hayashi, C. (1993). Nihon-jin no kokuminnsei (Japanese national character). *Phase, 93,* 64–96.
2. Nakamura, T., Yoshino, R., Maeda, T., Inagaki, Y., & Shibai, K. (2017). A Study of the Japanese National Character: The Thirteenth Nationwide Survey (2013)—English Edition. *ISM Survey Research Report,* No. 119. The Institute of Statistical Mathematics.
3. Hayashi, C., & Kuroda, Y. (1997). *Japanese culture in comparative perspective.* London: Praeger.
4. Yoshino, R., & Hayashi, C. (2002). An overview of cultural link analysis of national character. *Behaviormetrika, 29,* 125–142.
5. Forest Environment Study Group. (1981). International comparisons of attitudes toward nature. *Toyota Foundation Grant Research Report.*
6. Hayashi, C., Suzuki, T., Murakami, M., Yoshino, R., Hayashi, F., Kamano, S., Miyake, I., & Sasaki, M. (1988). *Kokuminsei nanakakoku hikaku [Comparison of seven countries' national character].* Tokyo: Idemitsu-syoten.
7. Zheng, Y., Yoshino, R., & Murakami, M. (2006). The sense of natural environments: A comparative study of Japan and China. *the Japanese Journal of Behaviormetrika, 32*(2), 55–68.
8. Yamaoka, K., & Hayashi, C. (2003). Cultural link analysis of Japanese Americans. *Behaviormetrika, 30,* 7–19.
9. Miyamoto, F. S., Fugita, S. S., & Kashima, T. (2002). A theory of interpersonal relations for cross cultural studies. *Behaviormetrika, 29*(2), 149–184.
10. Fugita, S. S., Kashima, T., & Miyamoto, F. S.(2002). Methodology of comparative studies of national character. *Behaviormetrika, 29*(2), 143–148.
11. Fugita, S. S., Miyamoto, F. S., & Kashima, T. (2002). Interpersonal style and Japanese American organizational involvement. *Behaviormetrika, 29*(2), 143–148.
12. Yoshino, R. (Ed.). (2000). *A study of statistical science on cultural transmission: Japanese Americans on the West Coast Survey (JAWCS). ISM Research Report 84.* Tokyo: The Institute of Statistical Mathematics.
13. Rotter, J. B. (1971). Generalized expectations for interpersonal trust. *American Psychologist, 26,* 443–452.
14. Uslaner, E. M. (2018). *The Oxford handbook if social and political trust.* Oxford University Press.
15. Yoshino, R. (2015). Trust of nations: Looking for more universal values for intrapersonal and international relationships. *Behaviormetrika, 42*(2), 131–166.
16. Yoshino, R., & Osaki, H. (2013). Subjective social class, sense of satisfaction, and sense of trust—a note on psychological scales of social surveys [in Japanese]. *Japanese Journal of Behaviormetrics,* 40, 2, 97–114. https://www.jstage.jst.go.jp/article/jbhmk/40/2/40_97/_article/-char/ja/, Accessed January 11 2016.
17. Rosenberg, M. (1956). *Misanthropy and political ideology* (pp. 690–695). XXI: American Sociological Review.
18. Almond, G. A., & Verba, S. (1963). *Civic culture.* Boston: Little Brown.
19. Robinson, J. P., Shaver, P. R., & Wrightsman, L. S. (Eds.). (1991). *Measures of social psychological attitudes* (Vol. 1). Measures of personality and social psychological attitudes: Academic Press.
20. Kawachi, I., Kennedy, B. P., & Wilkinson, R. G. (1999). *Income Inequality and Health: A Reader*. New York: The New Press.
21. Uslaner, E. M. (2002). *The moral foundations of trust.* Cambridge University Press.
22. Yoshino, R. (2002). A time to trust—a study on peoples' sense of trust from a viewpoint of cross-national and longitudinal study on national character. *Behaviormetrika, 29*(2), 231–260.
23. Inagaki, Y., & Maeda, T. (2015). An investigation of meanings of "trust" and their transition latent structure analysis on the meaning of sense of trust and longitudinal changes on the Japanese National Character Survey (in Japanese). *Proceedings of the Institute of Statistical Mathematics, 63*(2), 277–297.

24. Yoshino, R., & Tsunoda, H. (2010). A note on social capital–from a viewpoint of cross-national comparative methodology [in Japanese]. *Japanese Journal of Behaviormetrics, 37*(1), 3–17.
25. Fukuyama, F. (1995). *Trust*. NY: Free Press.
26. Dogan, M.(2000). Deficit of confidence within European democracies. In M. Halller (ed.), *The making of the European Union* (pp. 243–261). Oaris: Springer.
27. Yoshino, R. (2014). Trust of nations on cultural manifold analysis (CULMAN): Sense of trust in our longitudinal and cross-national surveys of national character. In M. Sasaki (ed.), *Cross-National Studies on Sense of Trust,* Chap. 7, Tokyo: Chuo-University Press.
28. Hofstead, G., Hofstead, G. J., & Minkov, M. (2010). *Culture and organizations—software of the mind* (3rd ed.). NY: McGraw-Hill.
29. Yoshino, R. (2009). Reconstruction of trust on a cultural manifold: Sense of trust in longitudinal and cross-national surveys on national character. *Behaviormetrika, 36*(2), 115–147.
30. Yoshino, R., Hayashi, F., & Yamaoka, K. *Kokkusai-hikaku data no kaiseki (Analysis of cross-national survey data).* Tokyo: Asakura-syoten.
31. Yoshino, R. (2005). A time to trust in the East Asia—a Behaviormetric study on the sense of trust in the East Asia Values Survey. *Japanese Journal of Behaviormetrics, 32*(2), 147–160.
32. Hayashi, F., & Nikaido, K. (2009). Religious faith and religious feeling in Japan: analyses of cross-cultural and longitudinal surveys. *Behaviormetrika, 36*(2), 167–180.
33. Zheng, Y., & Yoshino, R. (2003). Diversity patterns of attitudes toward nature and environment in Japan, USA, and European nations. *Behvaiormetrika, 30*(1), 21–37.
34. Reuters. (2013). China struggling to meet 2011–2015 environment goals (December 25, 2013). https://www.reuters.com/article/2013/12/25/china-environment-idUSL3N0K40YK20131225. Accessed March 14, 2015.
35. Reuters. (2015). China orders two local governments to punish polluting steel mills. (March 2, 2015). https://www.reuters.com/article/2015/03/02/us-china-steel-environment-idUSKBN0LY0C120150302.
36. Chen, Y. (2020). Research report on environmental changes in rural China –based on 2013–2019 fieldwork in Beicheng Village. *Bulletin of Data Analysis of Japanese Classification Society, 9*(1), 27–48.
37. Sasaki, M., & Suzuki, T. (2002). *Social attitudes in Japan: trends and crross-national perspectives.* Leiden: Brill Academic, Pub.
38. Yamamoto, K., Kawai, T., Wakisaka, K., Miyao, S., Mori, K., Hayashi, C., Midzuno, H., Suzuki, T., Hayashi, F., & Yoshino, R.(1993). Research on national character of Japanese Brazilian—1991–1992. *ISM Research Report,* No. 74.
39. Yamaoka, K. (2000). Variation in attitudes and values among Japanese Americans and Japanese Brazialians across generations. *Behaviormetrika, 27*, 125–151.
40. Kitano, H. H. L. (1993). *Generations and identities: tehe Japanese Americans.* MA: Ginn Press.
41. Seki, H. (2019). *Why are Confucious teachings "good" but Confucianism "bad"?* (in Japanese). Tokyo: PHP Institute.
42. Zheng, Y. (2005). Cross-national comparsion of transitions of traditional values in Eastern countries. *Japanese Journal of Behaviormetrics, 63*, 161–172.
43. Todd, E., & Garrioch, D. (1989). *The explanation of ideology: family structure and social systems.* NJ: Blackwel Pub.
44. Gallup, G. H. (1977). Human needs and satisfactions: a global survey. *Public Opinion Quarterly,* winter, 459–467.
45. Inkeles, A. (1997). *National character*. New Brunswick: Transaction Publishers.
46. Morishima, M. (1984). *Naze nihon ha seikou sitaka (Why did Japan succeed?).* Tokyo: TBS Britannica.
47. Dowds, M. (1986). *Buriea no kaihou-sya-tati* (Liberation army of Briar). Bunsyunn-bunko. Tokyo: Bungeisyunjyuu.
48. Kashima, T. (2003). *Judgment without trial: Japanese American imprisonment during World War II (The Scott and Laurie Oki Series in Asian American Studies).* Seattle: University of Washington Press.

49. Fugita, S. S., & Fernandez, M. (2004). *Altered lives, enduring community: Japanese Americans remember their World War II incarceration (The Scott & Laurie Oki Series in Asian American Studies).* Seattle: University of Washington Press.
50. Yoshino, R., Shibai, K., & Nikaido, K. (eds.). (2015). Sense of Trust—Summary Report: The Asia-Pacific Values Survey—Cultural Manifold Analysis(CULMAN) on People's. *ISM Survey Research Report*, No. 117.

Chapter 5
Further Development of CULMAN: Expanding of Comparability

This chapter presents some attempts to expand cross-national comparability on the paradigm of CULMAN. It covers the reduction of bias due to general response tendencies, the expansion of comparability across time in survey, the universality of social values, the resolution of essential differences between Eastern and Western religions, and international political relations. The examples are tentative, but they will give you some ideas as to how you can consider the methodology of cross-national comparison. Finally, an important comment is given on the impact of the survey cooperation rates in the analysis of survey data.

5.1 Re-Categorizations of Responses to Reduce Differences of General Response Tendencies

First, what cannot be ignored is the general response tendency of the people of each country (Sect. 3.4 of Chap. 3). For example, Japanese tend to avoid extreme answers and choose vague or medium answers. Americans and Arabs answers yes/no clearly, and French tend to respond negatively or critically. Japan reached the top class of the world economy in the 1980s, but their life satisfaction did not appear high in attitude surveys, which made the world wonder. Although Brazil was the world's largest debtor in the late 1980s, people's satisfaction and well-being appeared to the level as high as that of a wealthy country in Europe, which also seemed paradoxical [1].

Self-evaluation of people's life satisfaction and well-being can change significantly even during one day. But in the long run, it remains stable despite changing conditions (Kahneman and Krueger [29]). The long-term tendency seems to be closely related to the personality of each individual, country, and ethnic group. It may not seem to depend sensitively on the improvement or deterioration of the external environment such as politics and economy (Hofstead et al. [2], Chap. 6). In short, people's "response distribution" in attitudes surveys is a composite of reflection of

R. Yoshino, *Cultural Manifold Analysis on National Character*, Behaviormetrics: Quantitative Approaches to Human Behavior 10,
https://doi.org/10.1007/978-981-16-1673-0_5

the actual situation and the general response tendency. This involves a lot of problems for cross-national comparability of survey data. Let's start with the relatively simple attempt to ensure comparability.

First of all, response categories are often coarsened by re-categorization as an easy way to reduce differences in the general response tendencies for comparability between countries. For example, when asking about life satisfaction, the list of response choices is often in the form of "1. Very satisfied, 2. Satisfied, 3. Slightly dissatisfied, 4. Very dissatisfied," or the like. (Sometimes a neutral choice such as "Depends on" may be inserted between "2" and "3".) After data collection, we may re-categorize the positive options ("1" and "2") and the negative options ("3"and "4") as "1. Satisfaction" and "2. Dissatisfaction," respectively. This may be expected to reduce, for example, the differences between the Japanese and Americans regarding general response trends.

However, in some cases other re-categorization may be appropriate. Yoshino and Osaki [3] analyze the distribution of responses on "satisfaction" from all past major cross-national surveys of the Institute of Statistical Mathematics (see file:///C:/Users/yoshino/Downloads/kenripo117.pdf, pp. 103–143). Because the choices of question are at most on an ordinal scale, the meaningful statistical representative value of each population is *mode* or *median*, rather than *arithmetic mean* [4]. The past surveys show that the mode or median of responses to the satisfaction question is "1. Very Satisfied" only for Indians and Japanese-Brazilians but "2. Slightly Satisfied" for people of all the other countries/regions. Therefore, it may be better to re-categorize as "very satisfied (1)," "satisfied" (2), and "dissatisfied" (3, 4, and 5). Or, in some countries, it may be better to classify them as "very satisfied (1)" and "others" (2, 3, 4, 5).

People's life satisfaction has a high correlation with their satisfaction with every aspect of life such as family, health, and environment [5]. On the other hand, people's life dissatisfaction exhibits various patterns depending on the matter, as Tolstoy wrote in "Anna Karenina": "Happy families are all alike, every unhappy family is unhappy in its own way."

Incidentally, response distribution of "anxiety" also seems to be fairly varied on the country/region, gender, and age, depending on the theme such as serious illness, natural disaster, nuclear facility, traffic accident, and war. The social conditions and individual personalities of each country /region are intricately intertwined [3].

5.2 Connection Between Overlapping Local Charts Across Time Differences

A key concept of CULMAN entails the connection of two overlapping local charts. A pair of partially overlapping charts may form a larger chart if they are adequately connected to guarantee comparability (or common scalability) of all members of the connected charts. Aspects of comparability to be considered are: (1) comparability across two partially overlapping spatial local charts, (2) comparability over time of two temporal local charts, and (3) comparability of local charts of question items on a common theme. Here, let us reconsider an illustrative example of Sect. 4.2.5 in Chap. 4 on spatial and temporal connections between two local charts based on an item of two Asian social values versus two Western social values Table 5.1, as provided below [6, 7].

Table 5.1 (the upper part) shows data from the Seven-Country Survey (Japan, the United States, and five European countries surveyed during 1987–1993), and Table 5.1 (the lower part) from EAVS (Japan, China [Beijing, Shanghai, Kunming, Hangzhou, and Hong Kong], Taiwan, South Korea, and Singapore, surveyed during 2002–2005). Japan was surveyed in 1988 and 2002, and the response patterns seem fairly similar in both surveys. If it can be assumed that peoples' response patterns to the question remain relatively stable over the years, it may be justifiable to use

Table 5.1 Asian values versus Western values. This table shows response distributions in the Seven Country survey and the EAVS to Q.33: "If you were asked to choose the two most important items listed on the card, which two would you choose? (The choices are a,b,c and d, shown at the bottom of this table.)

		Par and Ben	Par and Right	Par and Free	Ben and Right	Ben and Free	Right and Free
Seven Country Survey (1987–1993)	Italy	23.7	**29.2**	26.3	4.4	2.5	14.0
	France	19.3	13.7	20.4	8.4	11.4	**26.8**
	West Germany	8.8	28.0	19.3	3.6	3.4	**37.0**
	The Netherlands	11.6	**30.1**	26.3	2.7	1.1	28.2
	UK	**31.8**	20.4	12.5	11.3	8.0	16.1
	USA	19.5	**39.4**	12.5	6.3	2.8	19.6
	Japan (1988)	**47.4**	13.7	19.5	4.8	6.9	7.8
EAVS (2002–2005)	Japan (2002)	**47.2**	13.5	18.8	4.8	8.1	7.6
	Beijing	**51.0**	21.8	14.9	3.8	1.8	6.6
	Shanghai	**51.5**	25.7	12.3	3.3	1.3	5.8
	Hong Kong	**35.6**	16.3	29.1	4.1	5.7	9.2
	Kunming	**49.8**	20.6	17.1	4.4	2.9	5.1
	Hangzhou	**48.6**	22.5	16.1	4.5	2.6	5.8
	Taiwan	**56.7**	19.5	14.2	3.2	2.8	3.6
	S. Korea	**45.2**	21.7	19.2	4.0	3.4	6.5
	Singapore	**50.3**	25.5	15.2	2.9	1.2	4.8

Filial piety/love and respect for parents
Repaying people who have helped you in the past
Respect for the rights of the individual
Respect for the freedom of the individual"
The choices are abbreviated as "Par," "Ben," "Right" and "Free," respectively, in this table

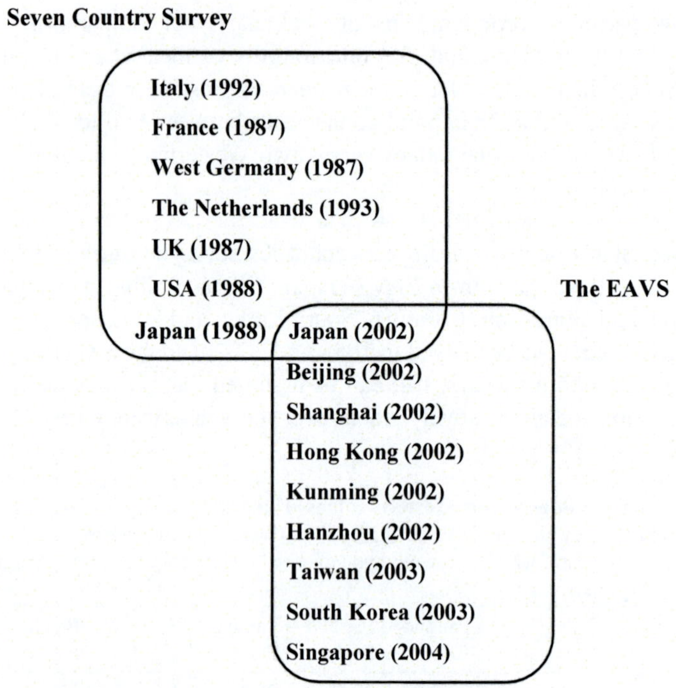

Fig. 5.1 Connection of two local charts: the Seven Country Survey (1987–1993) and the EAVS (2002–2005). Japan showed almost the same response distributions in both of the surveys conducted in 1988 and 2002 with respect to the item (see Table 5.1). These two local charts may be integrated into a single larger local chart to expand cross-national comparability beyond survey year differences, based on the assumption that the response distribution on the item is stable over time in each country.

Japan as a kind of connector between these two charts (the Seven-Country Chart and the EAVS chart). This means an expansion of comparability of the countries/area of the two charts beyond the differences in survey years (see Fig. 5.1). Of course, such an assumption may not be always completely justifiable. But, unless there's no significant counter-evidence, it may be expected to be useful for tentative analysis from a broader view. It is a kind of "abductive inference." For interpretations or justifications, it would be helpful to reconsider the arguments of Sect. 4.2.5 of Chap. 4 on universal social values.

5.3 Universality of Values

Chapter 4 examined a comparison of APVS countries based on items related to the teachings of Confucius, the choice of Asian and Western values, and the importance of some aspects of daily life (Tables 4.7, 4.8, and 4.9). The Confucius teachings originated in ancient China and were propagated in other Asian countries, including the Korean peninsula. It is not surprising even if these teachings, due to their historical origin, are not considered culturally significant in the Western world. However, responses in the United States, for example, are more positive regarding some of the items than those in Asian countries. Moreover, regarding the four choices provided for the item in Table 4.9, the most commonly selected choice in almost every country was "love and respect for parents" (Q34 of APVS) in both Asian and Western countries. Regarding the evaluation of the importance of an item in daily life (Table 4.8), the importance of family (Q 18 of APVS) scored highest in all the countries. The data suggest that "love and respect for parents" and "importance of family" may be universal social values beyond cultural differences. This can be an example where certain questionnaire items cover universal features, although there are many items to emphasize the uniqueness of a culture, country, or area.

5.4 Resolution of Intrinsic Difference between Eastern and Western Religions

Another example concerns religion. The Japanese word for religion is *Shuu-Kyo.* This word was coined during the Meiji Restoration Period (about 1867–1899) as a combination of the words *Shuu* and *Kyo. Shuu* means something that is essential or universal beyond the verbal or pictorial explanation, but some religious denominations have used certain characters or icons as "*mandala*" for the general public to understand it. *Kyo* means teachings. During that period when many new technologies and concepts were being imported into Japan, the Japanese occasionally needed to coin new words for new concepts or things from the western world. We recognize the difference in the concept of "religion" between Japanese and Westerners. Some Western researchers claim that Japanese and Asian religions such as Buddhism and Hinduism are not at all religious. Droit [8] argues that Westerners fear the existence of Japanese Buddhism and other religions that recognize life in mountains, rivers, trees and plants and praise "nothingness (the concept that everything is virtual existence)."

JNCS shows that some 60–70% of Japanese think that religious attitudes (or "hearts/minds") are important, even though only one-third or less have religious faiths. Christians or Muslims would have some difficulty understanding this situation. The question that arises is: why don't some Japanese individuals have a religious faith even though they consider religious attitudes (hearts/minds) to be important? The answer to this question may be closely related to the Japanese concept of *Shuu-kyo* which is more universal in some sense than the conception of Western or monotheist individuals, although the subtle differences between "religion" and *Shuu-kyo* can never be completely explained.

Christianity has the Bible, Islam has the Quran, and literally following it is faith. However, in Buddhism, for example, Sutra is a provisional teaching that respond to changes in the times and differences in countries and regions. And "*Shuu*" is essential

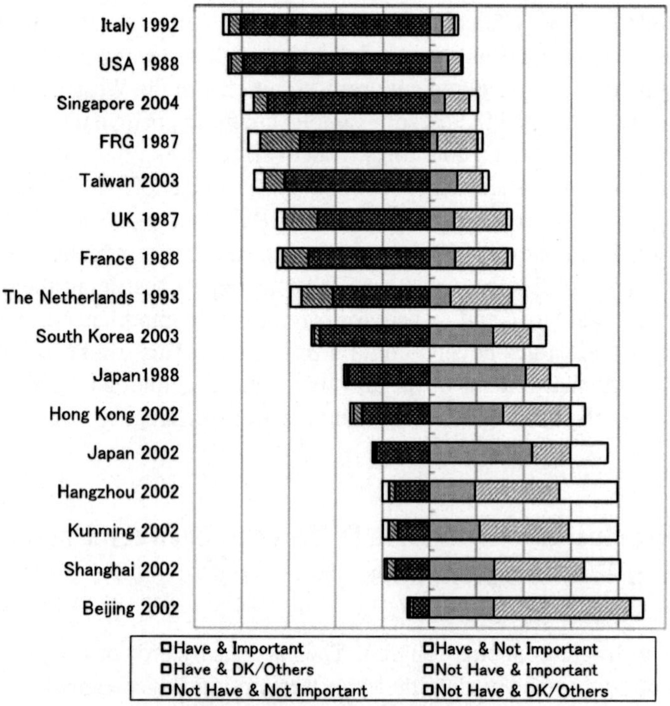
Italy 1992
USA 1988
Singapore 2004
FRG 1987
Taiwan 2003
UK 1987
France 1988
The Netherlands 1993
South Korea 2003
Japan1988
Hong Kong 2002
Japan 2002
Hangzhou 2002
Kunming 2002
Shanghai 2002
Beijing 2002
Have & Important
Have & Not Important
Have & DK/Others
Not Have & Important
Not Have & Not Important
Not Have & DK/Others

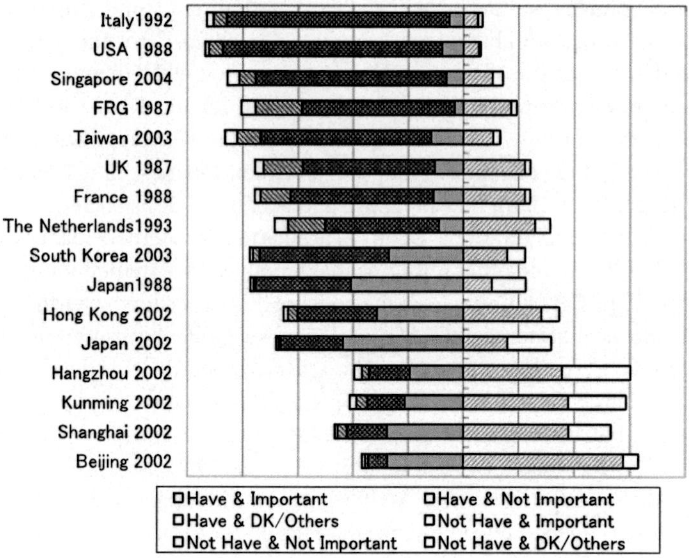
Italy1992
USA 1988
Singapore 2004
FRG 1987
Taiwan 2003
UK 1987
France 1988
The Netherlands1993
South Korea 2003
Japan1988
Hong Kong 2002
Japan 2002
Hangzhou 2002
Kunming 2002
Shanghai 2002
Beijing 2002
Have & Important
Have & Not Important
Have & DK/Others
Not Have & Important
Not Have & Not Important
Not Have & DK/Others

◂**Fig. 5.2 a. (the upper part) Cross-national comparison of distributions of people who have/do not have a religious faith (have vs. not have], and of those for whom the religious mind(heart) is/is not important (important vs. not important).** The left side of the figure shows percentages of people with a religious faith, whereas the right side shows those of people with no religious faith. There is a large discrepancy between Western and Asian countries. Source: Adapted with permission from Fumi Hayashi [30]. FRG stands for Federal Republic Germany (West Germany before unification). **b. (the lower part) Cross-national comparison of distribution of "people with a religious faith or those who think the religious mind/heart as important (have or important)" versus "the others (people with no religious faith or who think the religious mind/heart as unimportant (not have & not important)".** For this comparison, apart from mainland China, the discrepancy between Western and Asian countries decreases. This may be one way of comparing them using a more meaningful scale. Source: adapted, with permission, from Hayashi [30]. "DK" stands for "Don't know"

and universal, but it cannot be fully expressed in letters or pictures. Traditional Japanese Shinto presupposes the existence of 8 million gods (in fact, it means that gods dwell in everything). Even when worshiping a particular god or God, people do not stick to absolute values and teachings. You may find this attitude or way of thinking even among Japanese Christians. (Here it is necessary to distinguish between traditional Shinto and Meiji Shinto. The latter was politically invented by the state and people were forced to worship the Emperor as a god from the Meiji period until the end of World War II.)

Observing in this way, for Westerners (or believers of monotheism), "having a religious faith" and "thinking the religious mind important" may largely overlap. But polytheists like the Japanese can respect the "faith" of others, whether or not they have their own religious affiliation. So they can be generous with the religions of others. (See Park and Yoshino [9] for the Japanese essence of religious consciousness.)

Although it is not possible to take a deeper look at religious philosophy here, let me briefly mention the perspective of expanding cross-national comparability based on the above ideas.

In many of our past cross-national surveys, two questions on religion have been included as follows.

Q43(a) Do you have any personal religious faith?

(b) Without reference to any of the established religions, do you think.

a religious mind is important, or not?

In the English of Q43(b), "mind" is sometimes replaced with "attitude" or "heart/mind." The original Japanese word "*Shuu-kyo teki na kokoro*" (religious heart/mind) is difficult to translate into other languages because this involves the cultural differences of the religions mentioned above.

For example, our previous data show that a much higher proportion of Westerners (United States, France, West Germany, Italy, Netherlands) have religious beliefs compared to Asians. Figure 5.2a highlights the differences in the response patterns between Asia and Western world. In this figure, the differences are so great that a direct detailed comparison of the East and the West countries may not make much sense.

Regarding the cross-tabulation of questions Q43(a) and (b) for EAVS data, Fumi Hayashi [10, 30, 31] suggests a possible method for examining religious attitudes/minds in cross-national comparisons of Asian and Western countries in more meaningful way. This entails comparing those who have religions in Western countries (or monotheism) with those who have religions and/or those who think "a religious mind is important" (see Figs. 5.2b). That is, it suggests to intentionally confuse the definition of religious faith and religious mind/heart (i.e., merge them) and to compare the response patterns of those Asian and Western countries. This seems to present more meaningful comparison. Recall that the idea of Cultural Link Analysis suggests us not to directly compare totally different countries but to compare countries that have some common features. The approach of Hayashi [10] can be considered an attempt to find a more general basis for comparing religious attitudes by deliberately obscuring the detailed differences of notion of religion. This may not be the perfect approach, but it seems to give us some important insights into the differences between religion and *Shuu-kyo* (or monotheism and polytheism).

Incidentally, apart from the issue on religious concept, in some countries like Vietnam, there is a strict distinction between "personal faith" and "religions (designated by the government)." In some countries, certain political control over religion results from experiencing political turmoil by religious groups in their long history. In Sect. 4.2 of Chap. 4, I have explained that in many countries, people are cautious about religious groups, even when they respect the religion itself.

5.5 QMIII on GSS 3 Questions of Trust

Up to the previous section, we have discussed mostly the so-to-speak one-dimensional connection between pairs of local charts. Here, we consider a multidimensional structure of local charts that appear according to the degree of similarity between countries. Each local chart (a cluster of countries) is created according to the similarity between countries, and all charts form a hierarchical structure (manifold) of the similarity between charts. Thus the manifold depends on the range of question items and the range of countries to be compared.

As an example, we apply multidimensional data analysis (QMIII) to the cross-national data for the three GSS questions related to trust analyzed in Sect. 4.2 of Chap. 4. The main purpose of applying QMIII here is to analyze the mutual relations (similarity/dissimilarity) between countries in the multidimensional structure, rather than just to discuss the level of "trust" on the unidimensional distribution of responses for each question item (Q36, Q37, and Q38). Results can be represented as a manifold of local charts that correspond to the similarity of response patterns between countries.

The following survey data are dealt with here:

a. The Seven Country Survey (1987–1993) [11].
b. The East Asian Value Survey (EAVS) (1992–1995) [12].
c. The Pacific-Rim Value Survey (PRVS) (2004–2009) [13].
d. The Asia-Pacific Value Survey (APVS) (2010–2014) [7]. (For details of data, see the website of ISM: https://www.ism.ac.jp/~yoshino/index_e.html).

Figure 5.3a, b, c, and d shows the results of applying QMIII to the three GSM questions. (I used SPSS ver.24.0, Optimal Scaling, which is mathematically equivalent to QMIII.)

In order to comprehensively consider the similarities/dissimilarities of politics, economy, culture of countries/regions or the like, I tentatively have created those local charts (circles surrounding countries in each figure). Each country is located in the QMIII output, based on the overall structure of the compared countries. Therefore, it should be noted that adding another country to compare sometimes can significantly change the appearance of the manifold.

However, based on past studies, the seven countries (Figurer 5.3a) often appear to show similar relationships not only in the analysis of trust items but also in the analysis of the other items. The similarities from geographic, historical, religious, and cultural backgrounds appear to form three clusters: Anglo-American cluster (USA and UK) and Latin cluster (French and Italian), and a geographically neighboring cluster (West Germany [as of 1987] and the Netherlands). Japan is far from the other countries, but it is relatively closer to the West Germany. Speaking of dimensions, the left and right sides along the first dimensional axis are divided into Latin and non-Latin, and the upper and lower sides along the second dimensional axis are divided into Japanese-German and Anglo-American. Table 4.4a–d of Sect. 4.2 in Chap. 4 showed the correlation coefficient patterns for each pair of two of the three GSS questions. Here, QMIII analyzes the structure of the data of three questions simultaneously. You might see an entire country structure that the analysis in Table 4.4 cannot directly explain.

QMIII shows the similarities between countries/regions as the corresponding distances in the figure. Therefore, two patterns are considered to be the same if they match each other through symmetric transformation (upside down, or left–right flip). (For matching by rotation, you need to pay attention to the eigenvalues corresponding to the axes.) Many of these countries/regions have been repeatedly surveyed in our studies, allowing long-term comparisons. However, as mentioned above, selections of slightly different countries/regions can result in different QMIII outputs. (Next section shows some examples of this.)

EAVS (Fig. 5.3b) includes urban areas of the mainland China (Beijing, Shanghai and Hong Kong), as well as the inland Kunming and Hangzhou. Taiwan and Singapore are also included. These may be considered to belong to Greater China. The survey in Kunming and Hangzhou was limited to EAVS, while PRVS included the United States, Australia, and India. In addition, APVS included Vietnam also.

The QMII patterns of PRVS (Fig. 5.3c) and APVS (Fig. 5.3d) seem to be fairly consistent. On the other hand, EAVS (Fig. 5.3b) pattern is slightly different. Due to the large number of Greater China countries/regions selected in the study, the entire

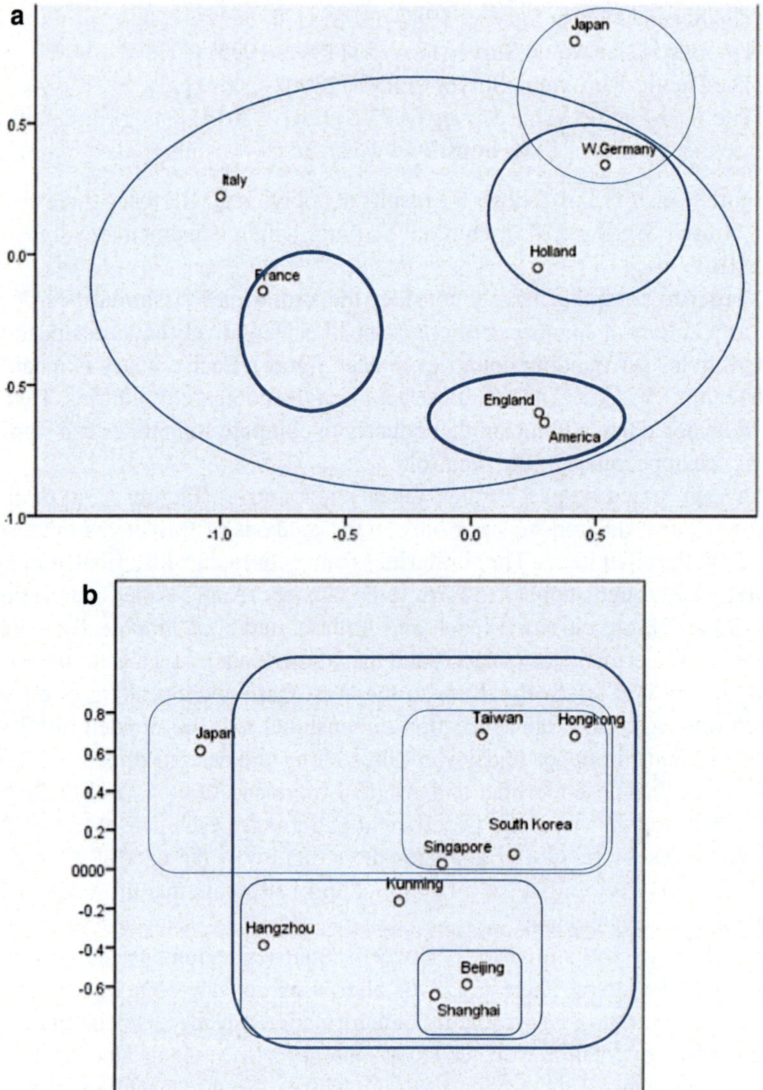

Fig. 5.3 **a Hayashi's Quantification Method III (MMIII) on GSS 3 Trust Items in Seven Country Survey: Japan, USA, UK, West Germany, France, Italy, and the Netherlands (Holland).** Eigenvalues of the first dimension and the second dimension are, respectively, 1.90 and 1.72. **b Quantification Method III (MMIII) on GSS 3 Trust Items in the EAVS: Japan, South Korea, Beijing, Shanghai, Konming, Hangzhou, Hong Kong, Taiwan and Singapore**. Eigenvalues of the first dimension and the second dimension are, respectively, 1.67 and 1.60. **c Quantification Method III (MMIII) on GSS 3 Trust Items in the PRVS: Japan, South Korea, Beijing, Shanghai, Hong Kong, Taiwan, Singapore, USA, Australia, and India.** Eigenvalues of the first dimension and the second dimension are, respectively, 1.67 and 1.57. **d Quantification Method III (MMIII) on GSS 3 Trust Items in the APVS: Japan, South Korea, Beijing, Shanghai, Hong Kong, Taiwan, Singapore, USA, Australia, India and Vietnam**. Eigenvalues of the first dimension and the second dimension are, respectively, 1.64 and 1.57

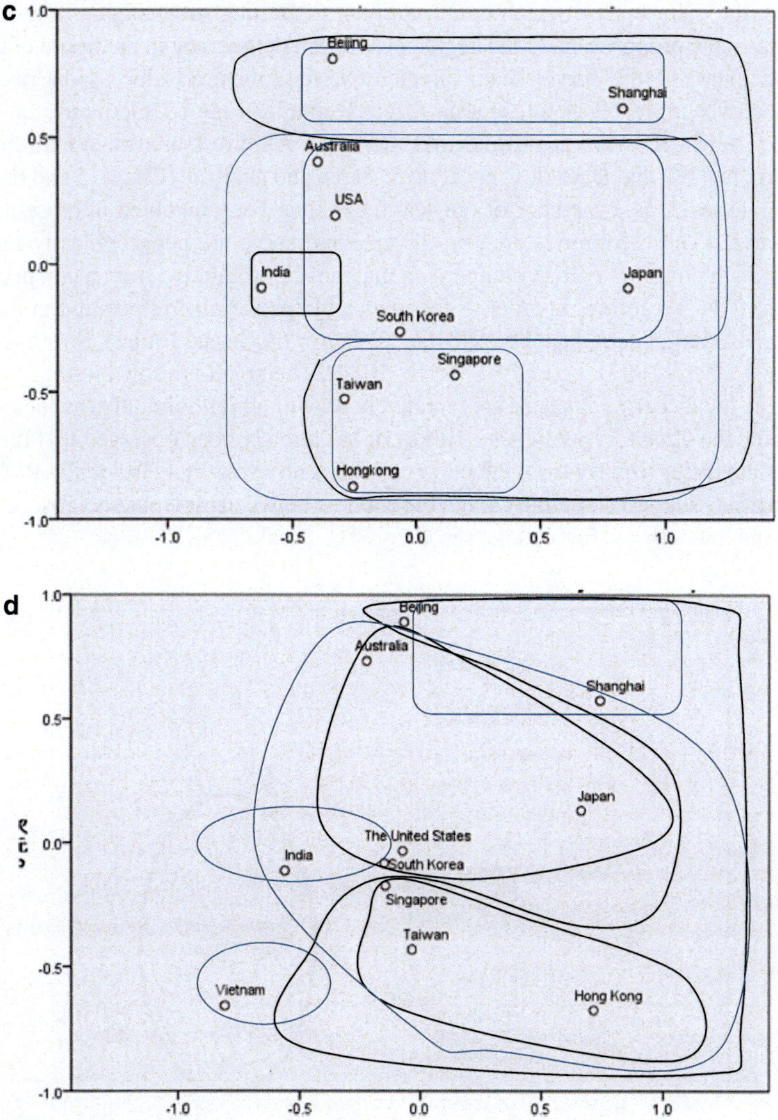

Fig. 5.3 (continued)

QMIII pattern of EAVS may be dragged into China (i.e., many countries/regions are located within the framework of China).

Nevertheless, these three figures clearly distinguish between Greater China and non-Greater China countries (charts), mainly in these Asian regions. And at the next level, the mainland of China and the neighboring countries (or non-democratic countries/regions and democratized countries/regions) are separated.

Also, the distance from each country/region to Beijing or Shanghai may seem to be inversely proportional to the degree of Western democracy in the figure of QMIII.

Incidentally, the survey team directed by Masamichi Sasaki [14] carried out a trust survey project targeting Russia, Czech Republic, Finland, Germany, the United States, Japan, Taiwan, and Turkey. As members, Akihiro Ishikawa, Vladimir Davidenko, and Nikolai Dryaklov report on research and analysis (Chaps. 5 and 6, Sasaki [15]). (Sasaki, as a member of our team, has long been involved in cross-national research.) These countries are very diverse and some are geographically far apart from each other, so a direct comparison may not be productive from a perspective of CULMAN. Therefore, in order to get an idea of the overall composition in line with CULMAN's concept, I applied QMII to all data of the Sasaki Project, Seven-Country Survey (1987–1993), and APVS (2010–2014). The result is shown in Fig. 5.4.

The local charts (clusters) are tentatively drawn to highlight only the locations of Russia, the Czech Republic, and Turkey. It has already been observed that India and Vietnam differ slightly from the other countries surveyed by PRVS and APVS. This figure may suggest something about the contrast between previous socialist countries and other countries. For details, see Chap. 7 of Sasaki [15].

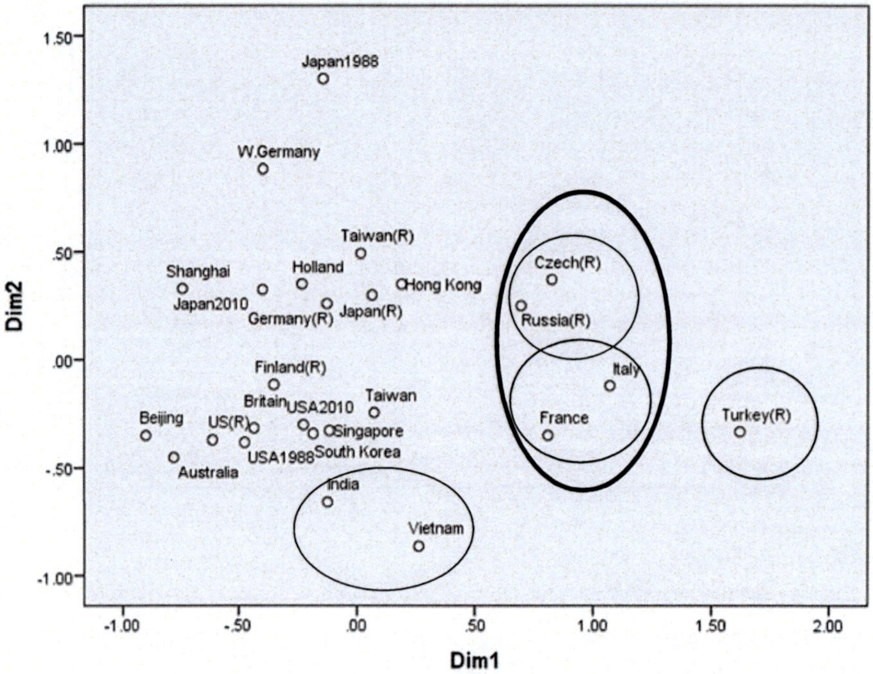

Fig. 5.4 Quantification Method III (MMIII) on the GSS 3 Trust Items applied to the data of the Seven Country Survey, the APVS and the Russian Survey Project (Russia, Czech, Finland, US, Japan, Taiwan, and Turkey Survey). *Note* (R) indicates part of a Russian research project by Sasaki [14]. Eigenvalues of the first dimension and the second dimension are, respectively, 1.64 and 1.57

5.6 Local Charts of International Political Relations

Fujita and Yoshino ([16], Figs. 8, 9, and 10) applied QMIII to the following sets of question items A, B, and C of the Pacific-Rim Values Survey (PRVS) to show the international relations of countries.

(A) Q2 (Country of friendship), Q3 (Country you want to be born), Q28 (Do you leave it to a good politician), Q31 (Individual priority or country priority), Q52e (Government trust), and Q52i (UN trust)
(B) Q28, Q31, Q52e, and Q52i
(C) Q28 and Q31.

Note that the set of question items above has an inclusive relationship of A ⊃ B ⊃ C. The exact wording of these questions is as follows. (PRVS and APVS used the same set of questions A, B, and C, but the question codes are different. Here I use the code of APVS.)

(Note: The choices of Q2 and Q3 exclude respondents' own countries and therefore vary by survey country. The USA version is shown below. Here [VOL] means that the interviewer does not read the choice aloud but records it only if the respondent voluntarily says something other than the choices offered.)

Q2 [SHOW CARD Q2] Which one of the following countries or regions would you like to see develop the friendliest relationship for our own national interest?

1 EU (European Union) 2 Japan 3 China (Mainland)

4 South Korea 5 India 6 Singapore 7 Australia

8 Other (VOL) ______________________[PLEASE SPECIFY]

9 DK

Q3 [SHOW CARD Q3] If you could be born again in a country or area other than the United States, which one of the following countries or area(s) would you like to be born in?

1. Japan 2. China (Mainland) 3. South Korea 4. Taiwan

5. Hong Kong 6. India 7. Singapore 8. Australia

88. Other (VOL) ______________________[PLEASE SPECIFY]

99. DK.

Q28 Some people say that if we get outstanding political leaders, the best way to improve the country is for the people to leave everything to them, rather than for the people to discuss things among themselves. Do you agree with this, or disagree?

1. Agree 2. Disagree 3. Undecided/it depends (VOL)

8. Other (VOL) ______________________[PLEASE SPECIFY]

9. DK

Q31 [SHOW CARD Q31] Please choose from among the following statements the one with which you agree most.

1. If individuals are made happy, then and only then will the country as a whole will improve

2. If the country as a whole improves, then and only then can individuals be made happy

3. Improving the country and making individuals happy are the same thing

8. Other(VOL)________________________________ [PLEASE SPECIFY]
9. Don't know

Q52 [**SHOW CARD Q52**] How much confidence do you have in the following? For each one, could you tell me how much confidence you have in them? Is it a great deal of confidence, quite a lot of confidence, not very much confidence or none at all? (READ OUT AND CODE ONE ANSWER FOR EACH)

	A great Deal	Quite A Lot	Not very Much	None At All	DK
e. Federal bureaucracy	1	2	3	4	9
i. The United Nations	1	2	3	4	9

(Note: Q52 has 10 items from a to j, but the above 2 items e and i were taken up.)

These items are obviously related to international relations, but it's worth mentioning Q3 and Q4. As a question about international relations, many researchers ask the respondents directly if they like a particular country, using a question format like Q3 (e.g., "Opinion Poll on Japan," Japanese Ministry of Foreign Affairs, https://www.mofa.go.jp/policy/culture/pr/index.html). In such a style of question, in many surveys, most of countries of the world show friendly attitudes for Japan whereas only China and South Korea keep anti-Japanese attitudes. However, if we ask the question in the style of Q4, even China and South Korea show more positive attitudes to Japan. For the question at EAVS, the first choice for South Korea was "Japan." Therefore, depending on the particular variation (wording) of the question, the respondent may reveal a very different attitude. This may be related to a sort of duality of the real mind and the superficial opinion (in Japanese, it is called "*Honne*" and "*Tatemae*" [cf. social desirability]). Therefore, it seems meaningful to include both Q3 and Q4 in international relations studies. PRVS did not include "Singapore" as a choice of Q2 and Q3 in some survey countries. Here we do the same analysis by Fujita and Yoshino [16], but use APVS data with "Singapore" as one of the choices for all countries surveyed (except Singapore itself).

The result for QMIII for the smallest set C of items is shown in Fig. 5.5a. Items in set C (Q28 and Q31) are directly related to the political system (e.g., individualism versus totalitarianism, oligarchy versus democracy). First, the distance between countries (similarities) seems to separate the United States and Australian from the other (Asian countries). Then, in the Asian group (local chart), the Indian sub-chart and the other Asian sub-charts look separated. Furthermore, the latter sub-chart may separate the group of Vietnam, South Korea, and Singapore from the group of Japan, Hong Kong, Taiwan, Beijing, and Shanghai. (These groupings are tentative and more substantive discussion and evidence may draw different local charts.)

For set B, the result is shown in Fig. 5.5b. Items in set B (Q28, Q31, Q52e, Q52i) are items of trust in government and the United Nations, in addition to the items in set C related to the political system. The right half is a group of democratically and economically developed countries and the left half is a country or area with an undemocratic system (or Greater China). It may be interesting that Hong

Kong is in the middle. The question that may arise is: do Hong Kong and India function as connectors between these two charts, or are they simply located in the middle because they are under transition, moving from one side to the other? Hong Kong has been under British rule for many years and has been unique since it was returned to China in 1998. Trust in government is an important issue, especially in the current relationship between Hong Kong's local government and Beijing, the central government. Its economic and political changes are drawing attention. India is also in the middle. It is the largest democratic country in the world in terms of population. However, the complex realities of multiethnic, multilingual, multireligious India must be carefully considered. Also, in India, trust in the government remains an important issue in improving democracy in a corrupt society. (International Transparency, NPO reports India has the highest ranking of Corruption Perceptions Index 2017 in Asia.) Corruption prevents foreign companies from investing in India (2017 A. T. Kearney Foreign Direct Investment Trust Index: https://www.kearney.com/foreign-direct-investment-confidence-index/2017-full-report). In 2018, the government enacted a new law called the Prevention of Corruption (Amendment) Act, 2018.

For set A, the result is as shown in Fig. 5.5c. In addition to the items in set B, the items in set A are the items of "Country of Friendship" and "Country to be Born." The bottom right shows the local chart of the United States, Australia, Japan, South Korea, Taiwan, Singapore, and Hong Kong. This local chart is overlapped to some extent by the local chart of India, Beijing, Shanghai, Hong Kong, and Singapore. Diagonally from the bottom right to the top left, the United States, Australia, Japan, South Korea, Taiwan, Singapore, Hong Kong, India, Shanghai, and Beijing are orderly arranged. It may be speculated that the diagonal line forms the axis of a substantial political system (e.g., corresponding to degree of democracy). Vietnam alone seems to form a unique chart of single country. Again, the position of India, the world's largest democracy, may be arguable. This figure makes us ponder about India's unique and complicated situation as manifold of casts (social classes) races, languages, religions, etc. Of course, the present Western democracy is not the only possible democracy. Even between European democracy and American democracy, we recognize some differences regarding the concept of human rights and freedom and their priorities. This may be a clue to thinking about possible democratic styles.

I leave it to the reader for a detailed discussion of substantive political and social issues. However, I will briefly explain the relationship between the sets of questions and the outcomes of QMIII.

As the sets of question items, the inclusion relationship $A \supset B \supset C$ holds. But the corresponding outcomes of QMIII show that the structure of the local charts is not always preserved according to the inclusion relation of the set of items. For example, set B does not necessarily correspond to a finer manifold holding the structure of the local charts of Set C. This means that certain additional items may create a more or less different manifold as an outcome of QMIII. (This also applies to adding another country.) In this sense, the effect of a set of items has a kind of gestalt (integrated wholeness). That is, "the whole is not a simple sum of all parts" (the main theme of Gestalt psychology).

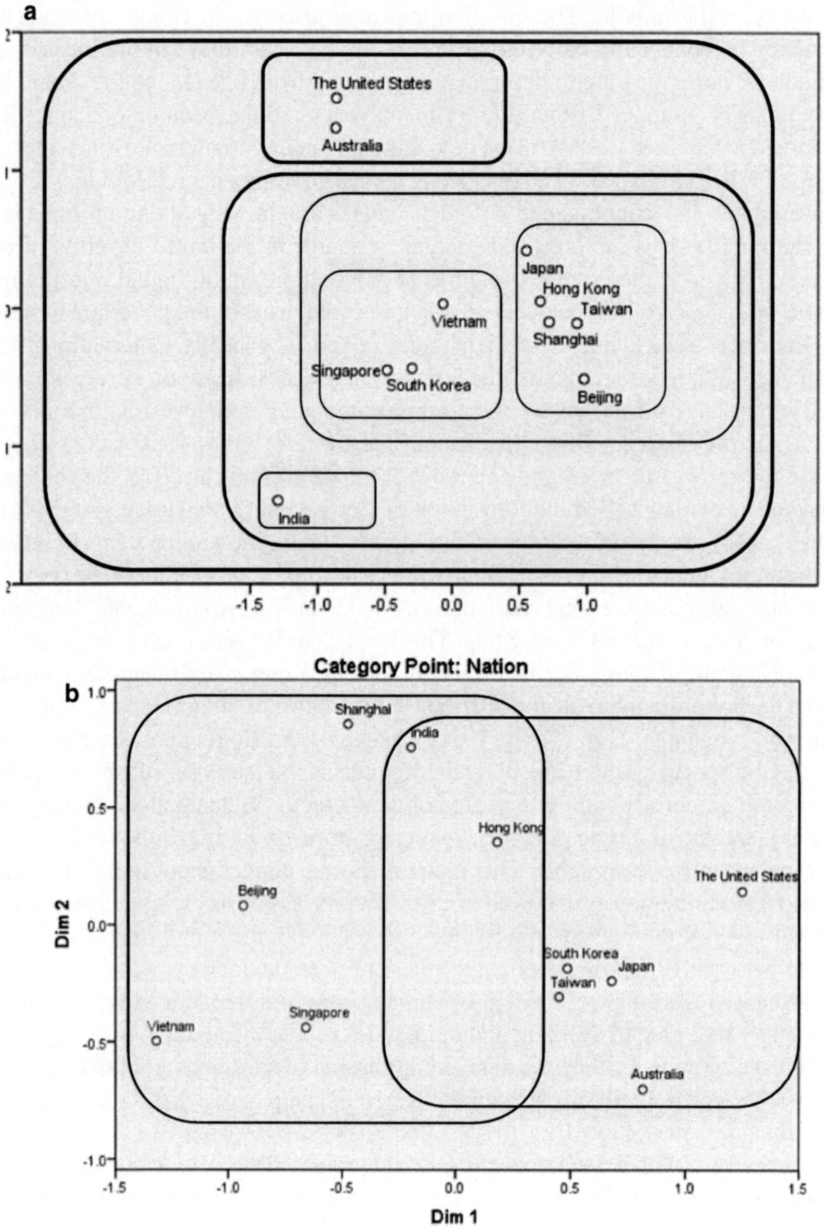
a
The United States
Australia
Japan
Hong Kong
Taiwan
Vietnam
Shanghai
Singapore
South Korea
Beijing
India
-1.5
-1.0
-0.5
0.0
0.5
1.0
Category Point: Nation
b
Shanghai
India
Hong Kong
The United States
Beijing
South Korea
Japan
Taiwan
Singapore
Vietnam
Australia
1.0
0.5
0.0
-0.5
-1.0
-1.5
-1.0
-0.5
0.0
0.5
1.0
1.5
Dim 2
Dim 1

◀**Fig. 5.5 a QMIII (Optimal Scaling of SPSS) on the data of Q28 and Q31**. A hierarchical manifold structure is evident here. The group of Asian countries and the group composed of the US and Australia constitute separate pairs of local charts. In the Asian chart, India is separated from the chart of the other countries. Furthermore, in the chart that does not include India, there is a pair of charts: one consists of Japan, Hong Kong, Taiwan, Beijing, and Shanghai, and the other consists of South Korea, Singapore, and Vietnam. The Eigenvalues of Axes 1 and 2 are 1.4 and 1.3, respectively. Q28: Some people say that if we get outstanding political leaders, the best way to improve the country is for the people to leave everything to them, rather than for the people to discuss things among themselves. Do you agree with this, or disagree? Q31: Please choose from among the following statements the one with which you agree most. 1. If individuals are made happy, then and only then will the country as a whole improve, 2. If the country as a whole improves, then and only then can individuals be made happy, or 3. Improving the country and making individuals happy are the same thing. **b QMII (Optimal Scaling of SPSS) on the data of Q28, Q31, Q50e, and Q50i**. The figure shows a pair of large overlapping charts. The chart on the right side consists of relatively developed democratic countries, whereas the chart on the left side consists of non-democratic countries (or the Chinese region). Hong Kong and India are located in the middle of the two overlapping charts. The Eigenvalues of Axes 1 and 2 are 1.9 and 1.6, respectively. See Figure 4a for Q28 and Q31. Q50: How much confidence do you have in the following? For each one, could you tell me how much confidence you have in them? How much confidence do you have in them? Are you very confident, somewhat confident, not confident or not confident at all? e. Government. i. the United Nations. **c QMIII (Optimal Scaling of SPSS) on the data of Q2 (Friendly Country), Q3 (Country to be born), Q28 (Political leader), Q31(Individual or Country), Q50e (Confidence in Government), and Q50i (Confidence in UN)**. The axis is formed diagonally from the lower right to the upper left, alongside the United States, Australia, Japan, South Korea, Taiwan, Singapore, Hong Kong, India, Shanghai, and Beijing. Only Vietnam seems to be off from other countries, the orthogonally projected component of Vietnam on this axis is close to India, Beijing, and Shanghai. The eigenvalues of the 1st and 2nd dimensions are 2.3 and 2.0

c

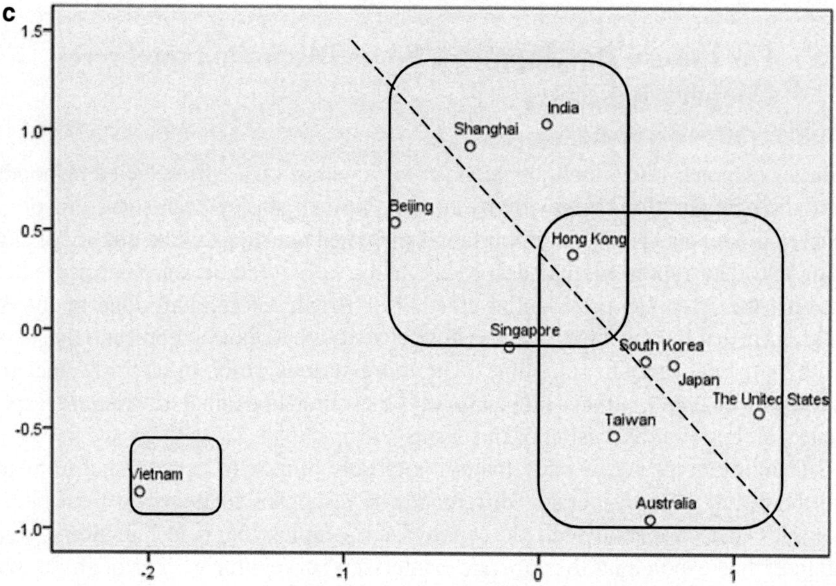

Fig. 5.5 (continued)

Another caution is needed. This kind of clustering (drawing local charts) cannot completely eliminate the arbitrariness of researchers. Researchers may draw different sets of local charts for the same outcome of QMII. To avoid this arbitrariness, for example, you can calculate the distance between countries from the multidimensional Cartesian coordinates (x_i, y_i, ..., z_i) for country i of the output of QMIII and apply certain clustering technique to the set of distance data. However, arbitrariness still cannot be eliminated completely, as there are various clustering techniques that give more or less different results. Therefore, regarding themes related to actual social problems, it seems more practical to proceed with data analysis while considering all accumulated empirical knowledge, rather than requiring automatic output of software. Superficial discussion should be avoided without considering complicated international relations and detailed circumstances of each country. But, at least, the following contrast seems clear: Western countries versus non-Western countries (individualism versus totalitarianism) in the figure of set C, democratic countries versus non-democratic countries (or Chinese cultural area) in the figure of set B, economically successful democratic countries and other countries (non-democratic or developing countries) in the figure of set A.

As to the issues that the set of local charts may change in relation to the set of items chosen, Matsumoto [32] and Fujita and Yoshino [16] have also provided examples that indicate a sort of trade-off between the range of countries studied and the depth of analysis regarding certain items such as people's sense of trust of social systems or institutions. (Here, the principles of complementarity should be noted again (Sect. 3.2 of Chap. 3).

5.7 For Future Development: Some Discussion on Survey Cooperation Rates

Survey response rates (including contact rates with statistically selected respondents and their cooperation rates) vary by country, time of survey year, survey institution, survey mode, etc. This is an important issue when considering the quality of survey data and data representativeness. Mass media polls made a serious misprediction against the 2016 US presidential election or Brexit referendum. One of the main reasons is that the response rate was not high enough to have a representative sample with statistical random sampling. (The more serious problem could be that those surveys used phone surveys or web surveys that could not calculate accurate response rates.) Below is a discussion of this issue.

The attitude of social participation is closely related to people's sense of trust, so there may be considerable differences of responses to items on trust between samples that cooperate with the survey and samples that reject it. Some studies suggest that respondents to cooperate in interview surveys have a high degree of social participation, such as community activities. Conversely, people non-cooperative in interview surveys, not always but often tend not to have smooth social life and tend to stay in their homes, dissatisfied with society, and not enjoying family life. This characteristic might overlap relatively with that of web survey respondents (Fushiki and Maeda [33]; [17–19]).

Let me show an example to highlight this problem. NHK (Japan Broadcasting Station) has been conducting the "Lifestyle and Attitude Survey of Junior High School and High School Students" since 1982, based on a rigorous statistical random sampling survey (NHK Broadcasting Culture Research Institute [34]). It was reported that the proportion of high school students who said they were happy had increased significantly in the past 30 years. The 2012 survey showed that "90% of high school students are happy." In a public symposium held on that theme, some education specialists disagreed with the data because they could hardly believe it really reflected the student situation ([34], Chap. 6).

Usually, the survey cooperation rate cannot be 100% in a rigorous statistical sampling survey. This does not necessarily impair the quality of the survey. The NHK uses strict random sampling, so we can calculate accurate response rates. Such calculations cannot be performed on poor quality data obtained, such as by quota sampling. However, when comparing random sampling survey data with very different response rates, special attention should be paid to the fact that superficial differences in response distribution can result from different response rates.

Very strictly, the true proportion of respondents who were "happy" in the total statistical sample is the proportion of the sum of the two groups over the total sample. That is, a group of respondents who answered "happy" in the survey, and a group of respondents who were "happy" but refused to cooperate in the survey. In general, the responses of those who refuse to cooperate in the survey cannot be observed. However, as mentioned before, it is speculated that people who are dissatisfied with their families, schools, and societies (i.e., unhappy people) may not tend to be cooperative in the survey.

Let's introduce some notations as follows:

- R_H the proportion of people who answered "happy" among the cooperated respondents.
- R the response rate (including rate of contact with respondents and their cooperation rate).
- N_H the number of people who answered "happy" in the survey.
- N_R the number of those cooperated in the survey.
- N_T the number of selected total sample.

Then, the following equations holds.

$$R_H \times R = (N_H/N_R) \times (N_R/N_T) \quad \ldots (1) \qquad = N_H/N_T \quad \ldots (2)$$

The second equation (2) is the proportion of people who answered "happy" among the entire sample (those cooperated and those not cooperated in the survey) selected from the population. (This immediately comes from the first equation (1), which is equal to the percentage of respondents who answered "happy" among those who cooperated in the survey multiplied by the percentage of cooperation in the survey.)

The longitudinal change of this value for high school students over the past three decades is as follows. The value of junior high school student is written in the parenthesis. (The second decimal place is rounded off below.)

2012: $87.9 \times 24 = 22.0$ (31.6).
2002: $86.4 \times 25 = 21.6$ (33.6).
1992: $81.1 \times 32 = 26.0$ (30.0).
1987: $74.5 \times 33 = 24.6$ (30.5).
1982: $63.4 \times 42 = 26.6$ (34.5).

(The survey used the leaving method in 2002 and 2012, and the mail-return method from 1982 to 1992. In the leaving method, the interviewer visited the respondents and asked them to complete the questionnaire by the day when the interviewer visit them again. In the mail-return method, the interviewer asks the respondent to fill out the answer and then return it to NHK. The response rates are fairly lower than usual NHK face-to-face interview surveys.)

Compared to the dramatic changes of "happy students" shown in the 2012 NHK report (from 63.4 to 87.9%), the multiplied values are rather stable in both high school and junior high school students. That is, of the overall sample, the proportion of students who cooperated in the survey and responded as "happy" has not changed so much over the last 30 years.

Although it is not possible to rule out all other possibilities from this data analysis alone, the rate of students who say that they are "happy" has not really changed much in the last 30 years.

On the other hand, a Web survey of the Ministry of Health, Labor and Welfare reported that the younger generation was less satisfied (Nikkei Shimbun electronic version July 13, 2014). As mentioned earlier, in web surveys, the sample of respondents seems to overlap with the "non-cooperative people" in the traditional mode (interview or mail) survey in Japan. A person who has a low self-disclosure tendency and is uncooperative with the interview survey may respond to a highly anonymous web survey and complains about social and family life [17, 18].

Actually, studying about "trust" over the past 20 years, I have a strong impression that the distribution of opinions is quite different between the people cooperated in the survey and those rejected to cooperate (Chap. 4 of this book, [20–26]). At least, the Japanese survey infers that in the survey of lower response rate, only the more "trustful" people cooperate in the survey. Therefore, the data of the lower response rate survey may look to yield the higher rate of respondents' "sense of trust." However, various effects of different survey institutions and survey methods are involved in each survey in a complicated manner, therefore, it is difficult to discuss quantitatively like NHK's happiness survey. (As for JNCS data, the trends of sense of trust show no large differences over 40 years even if the change of response rates are included. See Yoshino [26], Fig. 2.)

Overseas, it is not common for a well-developed list of residents to be available for public opinion polls, and the relationship between statistical random samples and total populations rarely can be completely defined by practical operations. Sampling procedures are too different to make exact cross-national comparison. However, if we attempt to calculate the Japanese-style strict survey response rate (considering contact rate, cooperation rate, etc.), many countries show as low as 20–40% [27]. This situation warns us about the representativeness of the survey data.

For example, although many studies on public health have shown a positive correlation between "well-beingness (physical, mental and spiritual health)" and "sense of trust" in each of many countries (Tsunoda et al. [35]), the correlation is not necessarily confirmed in cross-national studies. A possible interpretation is: perhaps the correlation might not be universal. Or, if the survey response rate is too low, we may be observing the data from the cooperated respondents who are likely to have higher sense of trust. If it is the case, because of the less variance of sense of trust, no correlation may be observed between well-beingness and trust (i.e., ceiling effect of variable on trust).

This section described how to integrate the survey response rate into the analysis of survey data. The survey data with a low response rate is often "corrected" for demographics such as gender, age, and ethnicity based on the census data. However, such corrections do not necessarily properly correct the response distribution of those who responded to the survey and those who did not, but often can increase unexpected bias. Such "correction" of bias is mathematically effective under the assumption that the response rate is high enough. But, then, the effect of the "correction" is about several percent at most. If the survey's response rate is too low such as less than 60%, then the assumption is not met and such "correction" can rather introduce unexpected bias. Thus, it seems much more important to scrutinize the original data, considering how the data is actually generated and what biases may be involved, rather than jumping to such a superficial numerical correction.

With this in mind, try to do some calculations on propensity scores, multiple imputations, calibrations, etc. Then find the question item that makes a big difference in comparing the original and modified data. It may provide some clues about total biases. Rather than superficial application of theory or technology, various trials and errors will lead us to the essential (see Note 1).

Note

Note1. The arguments of this section may be described in more general framework of survey cooperation. For example, reviewing various past studies on people's attitudes on survey participation, de Leeuw et al. [28] develop a short 9-item survey attitude scale that measures three important constructs related to decisions to participate in surveys: Interest in the survey theme, the value of the survey (social contribution), and the burden of participating in the survey.

5.8 Summary

This chapter exemplified some attempts to expand cross-national comparability based on the ideas of CLA and CULMAN. The simplest example is to loosely reclassify the categories of response data to reduce the bias of common response tendencies, such as the Japanese tending to avoid extreme choices and choose categories closer to the middle of choice. For some topics, if a particular country has shown a stable response pattern over the years in two sets of surveys, it may be possible to connect the two sets of countries as a set of comparable countries.

It is important and necessary to pay much attention to cultural differences when countries are compared across cultures, as superficially the same question may not mean the same for all people of different cultures. For example, there are essential religious differences that prevent a direct comparison of Eastern and Western religious attitudes. However, considering the religiosity of each culture at a deeper level can lead to a way to compare them. Similarly, the meaning of the same item on trust (such as the GSS items for trust) can vary from country to country, requiring a deeper level of data analysis for cross-country comparisons. On the other hand, in multidimensional data analysis such as QMIII, slightly different choices of questions about international politics may reveal significantly different aspects of the relationship between the target countries.

Finally, if the survey cooperation rates vary significantly between surveys, care should be taken when comparing survey data longitudinally or cross-nationally. The differences found between the two surveys or the two countries may not be due to the true difference, but to the difference in sample coverage.

References

1. Inkeles, A. (1997). *National character.* . New Brunswick: Transaction Publishers.
2. Hofstead, G., Hofstead, G. J., & Minkov, M. (2010). *Culture and organizations–software of the mind.* (3rd ed.). NY: McGraw-Hill.
3. Yoshino, R., & Osaki, H. (2013). Subjective social class, sense of satisfaction, and sense of trust–a note on psychological scales of social surveys [in Japanese]. *Japanese Journal of Behaviormetrics, 40*(2), 97–114. Retrieved January 11, 2016, from https://www.jstage.jst.go.jp/article/jbhmk/40/2/40_97/_article/-char/ja/.
4. Narens, L. (2002). *Theories of meaningfulness.* . London: LEA.
5. Yoshino, R. (2005). A time to trust in the East Asia–A Behaviormetric study on the sense of trust in the East Asia Values Survey. *Japanese Journal of Behaviormetrics, 32*(2), 147–160.
6. Sasaki, M., Yoshino, R., & Yano, Y. (Eds.). (2018). *Trust in modern society: Cross-national studies II.* . Tokyo: Chuo University Press.
7. Yoshino, R., Shibai, K., & Nikaido, K. (Eds.). (2015). Sense of trust-summary report: The Asia-Pacific values survey-cultural manifold analysis(CULMAN) on people's. *ISM Survey Research Report*, No. 117.
8. Droit, R.-P. (1997). *Le culte du neant: Les philosophies et le Buddha.* . Paris: Editions du Seuil.
9. Park, Y., & Yoshino, R. (2015). "Obake (Ghost) surveys" revealing underlying structure of heart and mind: Some relevant data from Asia Pacific Values Survey (APVS). *Proceedings of the Institute of Statistical Mathematics, 63*(1), 163–195.
10. Hayashi, F. (2013). The feeling of interpersonal trust and the view of religious mind-an international comparison. In Dryakhlov, N. I., Ishikawa, A., Kupreychenko, A. B., Sasaki, M., Toshchenko, Z. T., & Shadrikov, V. D. (Eds.), *Trust in society, business and organization: Conference proceedings of "business,society, human,"* (pp. 251–257). Moscow: National Research University.
11. Hayashi, C., Yoshino, R., Suzuki, T., Hayashi, F., Kamano, S., Miyake, I., Murakami, M., & Sasaki, M. (1998). *Kokuminsei nanaka-koku hikaku (cross-national comparison of seven nations).* . Tokyo: Idemitsu-Syoten.

12. Yoshino, R. (2008). Higashi ajia no kokuminnsei hikaku - data no kagaku [The cross-national comparison of the East Asia—science of data].
13. Yoshino, R., Hayashi, F., & Yamaoka, K. (2010). *Kokusai-hikaku deta no kaiseki (Analysis of comparative survey data).* . Tokyo: Asakura-Shoten.
14. Sasaki, M. (2012). Cross-national studies of trust among seven nationalities. In M. Sasaki & R. Marsh (Eds.), *Trust: comparative perspectives.* Brill.
15. Sasaki, M. (Ed.). (2014). *Cross-national studies of trust. Social Science Research Institute, Chuo University.* . Tokyo: Chuo University Press.
16. Fujita, T., & Yoshino, R. (2009). Social values on international relationships in the Asia-Pacific region. *Behaviormetrika, 36*(2), 148–165.
17. Hayashi, F. (2011). Survey on traditional Japanese Values and Daily Life (2)—Post Survey in Yokohama 2008. Toyoeiwa University. https://www.ism.ac.jp/~yoshino/other/dento/dento2.pdf.
18. Hayashi, F., & Yoshino, R. (2011). *Survey on traditional Japanese values and daily life—experimental surveys on comparisons of a postal survey and web surveys.* The Institute of Statistical Mathematics. https://www.ism.ac.jp/~yoshino/other/dento/dento.pdf.
19. Tsuchiya, T. (2010). Secondary adjustments of biases due to missing data by directional variables. *Proceedings of the Institute of Statistical Mathematics, 58*(1), 25–38.
20. Yoshino, R. (2009). Reconstruction of trust on a cultural manifold: Sense of trust in longitudinal and cross-national surveys on national character. *Behaviormetrika, 36*(2), 115–147.
21. Yoshino, R. (2012). Reconstruction of trust on a cultural manifold. In M. Sasaki & R. M. Marsh (Eds.), *Trust: Comparative Perspectives.* (pp. 297–346). Leiden: Brill Academic Publishers.
22. Yoshino, R. (2013). Trust of nations on cultural manifold analysis (CULMAN): Sense of trust in our longitudinal and cross-national surveys of national character. In Dryakhlov, N. I., Ishikawa, A., Kupreychenko, A. B., Sasaki, M., Toshchenko, Z. T., & Shadrikov, V. D. (Eds.), *Trust in Society, Business and Organization: Proceedings of the Conference, "Business, Society, Human"* (pp. 213–250). Moscow: National Research University.
23. Yoshino, R. (2014). Kouhuku-do ha seisaku-kagaku no tameni sokutei kanou ka? [Can "happiness" be measured as scientific scale for evidence-based policymaking? *Keikaku-gyosei* (Planning and Public Management), *37*(2), 35–40.
24. Yoshino, R. (2002). A time to trust–A study on peoples' sense of trust from a viewpoint of cross-national and longitudinal study on national character. *Behaviormetrika, 29*(2), 231–260.
25. Yoshino, R., & Tsunoda, H. (2010). A note on social capital–from a viewpoint of cross-national comparative methodology [in Japanese]. *Japanese Journal of Behaviormetrics., 37*(1), 3–17.
26. Yoshino, R. (2015). Trust of nations: Looking for more universal values for intrapersonal and international relationships. *Behaviormetrika, 42*(2), 131–166.
27. Yoshino, R. (2015). Report on the Asia-Pacifc Values Survey. *Conference Proceedings of the Japanese Society of Bahaviormetrics, 43*, 46–49.
28. De Leeuw, E., Hox, J., Silber, H., Struminskaya, B., & Vis, C. (2019). Development of an international survey attitude scale: Measurement equivalence, reliability, and predictive validity. *Measurement Instrument for the Social Sciences, 1*(9), 1–10. https://doi.org/10.1186/s42409-019-0012-x.
29. Kahneman, D., & Krueger, A. B. (2006). Developments in the measurement of subjective well-being. *Journal of Economic Perspectives, 20*(1), 3–24.
30. Hayashi, F. (2007). Reseacrh on religious faith and religious mind based on cross-national survey. IMPS 2007, abstract, p. 81. The 72nd Annual Meeting of the Psychometirc Society, Tokyo.
31. Hayashi, F. (2010). Contemporary Japanese religious mind - based on survey data from the Japanese national character and other cross-national surveys. *Proceedings of the Institute of Statistical Mathematics, 58*(1), 39–59.
32. Matsumoto, W. (2006). Sense of trust in organizations in East Asia: analysis for a cross-national comparative study. *Japanese Journal of Behaviormetrika, 33*(1), 25–40.
33. Fushiki, T., & Maeda, T. (2013). Adjustments on biases due to mssing data in recent social surveys (in Japanese). *Proceedings of the 42 annual conference of Behaviormetirics Scociety at Chiba University*, 236–237.

34. NHK Broadcasting Culture Research Insitutute. (2013). *NHK 2012 survey on junor and senior high school students' opinion on life - "happy" teenagers produced by the lost two decades (in Japanese).* Tokyo: NHK publisher.
35. Tsunoda, H., Yoshino, R., & Yokoyama, K. (2008). Components of social capital and socio-psychological factors that worsen the perceived-health of the Japanese males and females. *The Tohoku Journal of Experimental Medicine, 216*(2), 173–185.

Chapter 6
Conclusion: For Future Research

6.1 Survey Research for Mutual Understanding of the World

Statistical sampling survey data provides basic information for social research and evidence-based policymaking. Longitudinal and cross-national survey, among others, is a rich source of insights in scholars and social planning. However, as explained in the previous chapters, social survey data has various problems related to long-term comparability and cross-national comparability. To date, many methods of data collection and analysis have been invented and accumulated to overcome these problems. In Japan, many practical methods of statistical random sampling were established in the late1940s and 1950s, and over the past 30 years, many ideas have been presented for cross-national research. Many of these were achieved independently of the development of other countries, under special conditions for the restoration of post-World War II national power and the development of postwar democracy. Japanese researchers usually have no difficulty in integrating foreign research achievements published in English with their achievements. Unfortunately, however, Western researchers may find it difficult to access research and findings published in Japanese, and many may tend to ignore them.

Despite those past achievements, the issue of cross-country comparability has not been completely resolved. Longitudinal comparability of survey data also presents a difficult problem if the survey lasts for decades or more. However, as repeatedly emphasized in this book, each county's research method and research infrastructure itself reflect its own way of thinking, cultural, historical, religious, political, and social conditions.

Apart from the problem of practical research methodologies, there is more essential problem. Two countries may disagree within the framework of the same idea. Or two countries may show similar opinions on the surface, but within a completely different framework of thinking. People's values about life, religion, leadership, nature, etc. are not independent, but intricately mixed with each other. It is necessary

R. Yoshino, *Cultural Manifold Analysis on National Character*,
Behaviormetrics: Quantitative Approaches to Human Behavior 10,
https://doi.org/10.1007/978-981-16-1673-0_6

to deepen mutual understanding of the world by taking a cross-national comparative approach rather than simply comparing survey data superficially.

There may be national stereotypes. For example, the United States is a country created by Puritan and needs a grand design for the activities of all who continue to participate in the construction of the country and continue to monitor the authorities. In Russia, leaders continue to fight for power, people fear power and obey, and civil society is vulnerable. Diplomacy is for domestic politics and lies are normal. In China, central government executives are focusing on power struggles and domestic order, while local executives are seeking self-interest. People tend to be silent and pretend to obey the authorities, but exceeding the limits of patience can explode frustration. Japan is an adaptive country and does not have its own grand vision. This was often due to Prince Shotoku's administration around the six century. At that time, the Japanese witnessed a serious and tragic struggle for political power by those naturalized from the Korean Peninsula. After that, Prince Shotoku took the initiative to eliminate conflicts and emphasized the "harmony" of the people. Since then, the Japanese have been educated to pay more attention to others rather than sticking to self-assertion. As a result, the Japanese are tolerant of others and can actively adapt to others' ways. The Japanese, on the other hand, may appear to lack a spirit of independence and may not be liable for the consequences of important decisions or actions (i.e., vague responsibilities and lack of leaders to make clear decisions). For example, regarding the 2003 Iraq War, Japanese Prime Minister Koizumi declared that he would follow the US decision despite Japanese public opposition. After the war, both Britain and the United States admitted their mistakes, but Japan remained silent.

In future research,In it is important and necessary to continue to review these stereotypes and historical explanations in terms of substantive data.

6.2 Global Cultural Manifold

The last century was an era of expansion of Western civilization, and this century is said to be an era of reconstruction in Asia. Cultural differences often hinder our deep mutual understanding. In this era of globalization, world leaders must be sensitive to the social values of people if they want to take the responsibility of developing and maintaining world peace seriously. There are many ways to succeed in the development of society. Some institutional systems and customs have changed and are converging toward more universal under the influence of cross-border exchanges and trade. However, other systems are more sensitive to cultural differences in response to globalization. The last three decades have shown that, at least in the foreseeable future, globalization will not lead us to a single unified global culture or "superculture" on Earth. This is in line with the theory of cultural evolution that more variations increase the chances of survival. (For example, Inglehart [1], p. 42, presents a theory of cultural evolution that the value systems of different cultures have not converged but may be changing in the same direction with respect to "self-expression values.")

The principle of survival may suggest respecting multiculturism, but perhaps with some caution [2].

I think CULMAN can be utilized to develop a framework of policymaking for the gradual development of, so to speak, a global cultural manifold (GCM) ([3–8]) (see Fig. 3.7 of Chap. 3). GCM is a set of hierarchical local charts, with each local chart covering a specific area (region, country, group of countries, civilization, etc.). Local charts assume that people share a particular cultural or social value. Larger local charts are less restrictive but correspond to more universal cultural or social values. The entire set of local charts forms a kind of hierarchy. GCM charts dynamically change, so each chart may expand, split into two, or disappear over time. Some overlapping local charts may be assimilated and create a larger local chart. Sometimes a new local chart may appear.

Depending on the size of the local chart (area, region, or social group), people can create a set of rules for decision making or regulation of different types of exchanges (e.g., international trade within region members). For peaceful and steady integration and expansion of charts, a set of soft local rules to connect neighboring charts would be more effective, rather than a single strict global regulation. The set of local rules also may make a hierarchical structure with respect to its coverage and strictness.

The history of EU may exemplify the successful development of GCM, at least until the 2000s. But over the last two decades, some sort of over-integration, such as monetary union and the integration of economic policies that ignore states, has been dire.

In the eighteenth century, David Hume wrote in his essays [9] that "Nothing is more favourable to the rise of politeness and learning, than a number of neighboring and independent states, connected together by commerce and policy. The emulation, which naturally arises among those neighbouring states, is an obvious source of improvement: But what I would chiefly insist on is the stop, which such limited territories give both to power and to authority." As Hume wrote, the connection between neighboring independent states is important. The early developments of the EU were in line with Hume's ideas, but the last two decades seem to deviate from Hume's ideas. There seems to be a reason why Britain, the birthplace of Hume, has begun to leave the EU.

The key to successful global cooperation is a multifaceted approach to good connections with neighboring countries. Even if the two countries are competing with each other in military power, finding the benefits of close cooperation in the economy can limit their involvement in serious conflicts. Nay ([10], pp. 146–147) wrote that "As a preponderant power, the United States can provide an important public good by acting as a mediator." But, instead of imposing strict global regulations on the world by power, any country could be an intermediary between conflicting countries in each local region if it uses a set of soft rules for mediation. I think that the world leader in the twenty-first century is not a country that controls the world by military power, but a country that can act as a mediator of regional conflicts and form a world coalition. The most influential countries are those that can be mediators of regional conflicts around the world.

The most serious obstacles to successful international cooperation may appear to be political or religious fundamentalists who strongly believe in their own values and impose on others. Can there never be a way to connect them with people of different political ideologies and religions?

Occasionally extremists in monotheism may look insurmountable obstacles in connecting the world because they insist in dominating the world in their own way. For example, Islamic fundamentalism is said to never accept other religions. However, actual history shows that Islamic leaders of each era have led people to political adaptive behavior. On the one hand, this was the flexibility to seek the asylum of the ruler, and on the other hand, it was a visible deviation from the principles. Before the end of the Cold War, including Islam, people around the world usually tended to move away from religion. And the general public has lost more interest in religion and politics than ever before. However, the remarkable social disparity (the gap between the rich and the poor) was reflected in Islam in the form of religion = political extremism. It is also associated with European liberal modern values, and Islam in Europe has become much more radical than in the Middle East.

Historically, Islamic revelation texts were fixed, and the derivation of norms from revelations was completed in the tenth century. The legal scholars, who were obliged to continue to apply to the changing reality that followed, expanded the theory of law interpretation and devised practical treatments, and did not literally stick to old texts ([11], p. 173). Thus, except some minor extremists, fundamentalists can be realistic by their leaders and coexists with people of other religions.

A more recent issue of ethnic conflict is about immigrants in each country. Multiculturalism was originally aimed at alleviating racial and ethnic conflicts in each country. However, some argue that multiculturalism fosters conflict between migrants and the host country, as well as between migrants. For the people of the host country, immigrants appear to be over-treated by multicultural policies, and some argue that the relatively rapid increase in immigrants is a serious threat. For immigrants, multiculturalism reduced the economic gap between ethnic groups, but increased the gap within immigrants.

Can immigrants be a source of connections between different races? The answer depends on the attitude of immigrants and the national policies of each country. Immigration has been a source of American power for many years. In most developed countries, the average age is becoming higher and higher and the number of children is declining. But the United States is probably the only exception, as immigration continues. On the other hand, there was always friction between old and new immigrants.

The following can be considered for successful global collaboration.

1. Promote the growth of the middle class in each country and foster liberal and tolerant values.
2. Promote the formation of a chain of trust between countries.
3. Ordinally people are usually indifferent to politics and tend to leave it to their leaders. Therefore, leaders must always maintain the trust and responsibility of people, but people must continue to monitor the authorities.

Even if two countries are mutually exclusive (Local Chart A and Local Chart B), they may be able to be connected from A, A', A'', A''', ... to B by a chain of temporally changing or spatial neighboring charts. This means that instead of connecting the two countries directly, they can be connected by a series of changes of those countries or by forming an international chain of intermediaries.

6.3 Democracy

The end of the Cold War witnessed the democratization of many countries in Eastern Europe and South America, but some went to more turmoil for their own reasons. In addition, the last two decades in the EU and the United States have shown the turmoil of Western democracy, which has been a major model in the world for decades. People know that democracy does not automatically lead us to peace and prosperity. After World War II, in order to restore peace and economic power in a completely devastated country, the Japanese seriously considered the path to democratic development under the new constitution established during the American occupation. During times of hunger and poverty, Japanese Ministry of Education published a textbook of "democracy" for school children (Part I of 1948 and Part II of 1949) [12, 13]. The draft was written by many prominent Japanese scholars, including lawyers, economists and journalists. Finally, it was edited by Tomoo Odaka, along with General MacArthur's group (SCAP/GHQ, Chief of the Japanese Occupation Team). It was designated for primary school students and high school students, but the authors developed a deep discussion of political philosophy in a style that even young children could understand. The textbook reiterates that democracy does not automatically lead us to peace and prosperity, but that we must find ways to develop peace and prosperity under the democratic system.

Now is the time to rethink how world peace and prosperity can be developed within a democratic framework that embraces differences in culture, religion and social values. Perhaps the world of democracy can be created not by strict global standards, but by a set of soft rules that properly connect pairs of regionally diverse democratic systems.

On March 11, 2011, the Great East Japan earthquake caused a huge tsunami and resulted in the Fukushima nuclear power plant disaster. These brought seemingly insurmountable difficulties that Japan is still struggling to recover from. The world media, however, reported the calm attitudes of the Japanese even in the tragedy: the devastated yet surviving Japanese kept order in front of grocery stores to buy food, which may look remarkable to people of other countries where even a natural disaster can frequently lead to a riot. On the other hand, many Japanese have had a chance to reconsider the value of their own lives and work and to think of various ways of contributing to the people and area damaged by the quake. Many news stories and surveys have repeatedly reported on the human bond and the importance of family, relatives, and friends, not only on a domestic but also on a worldwide scale. We have confirmed that the differences in ideology or religions are minor compared to the universal importance of family, relatives, and friends, i.e., human bonds and trust between peoples.

It is my sincere hope that mutual understanding among the various cultures and civilizations will prevent serious conflicts between nations and cultures and will lead us to a peaceful and prosperous world in the twenty-first century.

References

1. Inglehart, R. D. (2018). *Cultural evolution: People's motivations are changing, and reshaping the world.* . London: Cambridge University Press.
2. Taylor, C. (1994). The politics of recognition. In A. Gutmann (Ed.), *Multicultureism* (Chapter 1). NJ: Princeton University Press.
3. Fujita, T. (2016). Examining people's attitudes and values relating to international relations in the Asia-Paicfic region. *Behaviormetrika, 43*(1), 41–63.
4. Fujita, T., & Yoshino, R. (2009). Social values on international relationships in the Asia-Pacific region. *Behaviormetrika, 36*(2), 148–165.
5. Yoshino, R. (Ed.). (2008). *Higashi ajia no kokuminnsei hikaku--de-ta no kagaku [The cross-national comparison of the East Asia--Science of Data --].* . Tokyo: Bensei-syuppan.
6. Yoshino, R. (2015). Trust of nations: looking for more universal values for intrapersonal and international relationships. *Behaviormetrika, 42*, 2, 131–166. Retrieved January 11, 2016, from https://www.jstage.jst.go.jp/article/bhmk/42/2/42_131/_article.
7. Yoshino, R. (2015). Cultural manifold analysis (CULMAN) as a paradigm of cross-national comparative surveys on national character. *Proceedings of the Institute of Statistical Mathematics, 63*(2), 203–228.
8. Yoshino, R., Shibai, K., Nikaido, K., & Fujita, T. (2015). The Asia-Pacific Values Survey 2010–2014–cultural manifold analysis (CULMAN) of national character. *Behaviormetrika, 42*(2), 99–130.
9. Miller, F. E. (Ed.). (1985). *Essays: Moral, political, and literary by David Hume.* . IN: Liberty Fund Inc.
10. Nay, J. (2002). *The paradox of American power.* Oxford University Press.
11. Iiyama, Y. (2019). *Islam 2.0* (in Japanese). Tokyo: Kawade-Shobo.
12. The Japanese Ministry of Education. (1948). *Democracy, Part I: textbook.* (Part I & II are republished as a book by Kadokawa Sophia-Bunko, 2018).
13. The Japanese Ministry of Education. (1949). *Democracy, Part II*: textbook.